I0487870

Better Than Kyoto

Better Than Kyoto

How Climate Stability Bonds can inject market incentives into the achievement of a stable climate

Ronnie Horesh

Writers Club Press
San Jose New York Lincoln Shanghai

Better Than Kyoto
How Climate Stability Bonds can inject market incentives into the achievement of a stable climate

All Rights Reserved © 2001 by Ronnie Horesh

No part of this book may be reproduced or transmitted in any form or by any means, graphic, electronic, or mechanical, including photocopying, recording, taping, or by any information storage retrieval system, without the permission in writing from the publisher.

Writers Club Press
an imprint of iUniverse, Inc.

For information address:
iUniverse, Inc.
5220 S. 16th St., Suite 200
Lincoln, NE 68512
www.iuniverse.com

ISBN: 0-595-21164-X

Printed in the United States of America

Introduction

The evidence that the global climate is changing is growing. But it is not conclusive and many questions remain unanswered. How fast is the climate changing? What will be the effects of climate change? Is there definitive proof that human activities are altering the climate? How much can we do about it? How much should do about it? We don't have definitive answers to these questions but even so, climate change has the potential to inflict serious harm on human, animal and plant life, so there is a strong case for doing what we can to minimise its adverse effects. The key intergovernmental body that was set up to assess the problem of climate change is the Intergovernmental Panel on Climate Change (IPCC) formed in 1988, cosponsored by the United Nations Environment Programme and the World Meteorological Organization and made up of over 2000 scientific and technical experts from around the world. Its deliberations and negotiations have culminated in the Kyoto Protocol ('Kyoto'), which is intended as a first step along the road to solving the climate change problem.

Chapter 1 of this book will briefly describe climate change and look at the major flaws of the Kyoto Protocol. Chapter 2 introduces a new financial instrument, Climate Stability Bonds, aimed at achieving climate stability more efficiently than Kyoto. It does this by channelling market forces into the solution of the climate change problem. Markets are the most efficient means yet discovered of allocating society's limited resources, but many believe that market forces inevitably accentuate extremes of wealth and poverty, and have little to do with managing the environment. It is true that deregulation and the freer operation of self-interest in parts of the private sector have made many individuals very wealthy indeed, while the less well off have gained little, the environment continues to suffer, and many social objectives remain as remote as ever. It is therefore important to remind ourselves at the outset that a market economy is consistent with many different outcomes. Much depends on the rules of the game: in this sphere the state and society have roles, both within a country and on a global scale. The market is one institution among many, and *market forces can be made to serve public, as well as private, goals.*

Climate Stability Bonds are intended to channel the market's incentives and efficiencies into the achievement of what is probably *society's* overriding environmental objectives: the achievement of a stable climate. After the brief description of the Bond concept in chapter 2, subsequent chapters look, respectively, at its advantages over Kyoto and practical aspects of its application. The fifth and final chapter looks at Bond principle's possibilities and limitations. The epilogue looks at people's response to the Bond concept, and at how it might be used to combat international terrorism, while the Annex presents a short paper on applying the concept to other social problems.

Chapter 1

Climate change and Kyoto: is the climate changing?

Weather can change very rapidly from day to day and from year to year, even within an unchanging climate. These changes involve shifts in, for example, temperatures, precipitation, winds, and clouds. Climate is more long term: essentially it is the average weather, including seasonal extremes and variations. Climate is controlled by the long-term energy balance of the Earth and its atmosphere, and is influenced by slow changes in such features like the ocean, the land, the orbit of the Earth about the sun, and the energy output of the sun.

The four warmest years on record since 1860 have all occurred since 1990, and the evidence certainly suggests that the planet is warming. The globally averaged temperature of the air at the Earth's surface is generally accepted as having increased between 0.3 and 0.6°C since the late nineteenth century.

The box presents excerpts from the *IPCC's Summary for Policymakers*, which was approved by IPCC member governments in Shanghai in January 2001. They describe the current state of understanding of parts of the climate system.

Excerpts from the IPCC's Summary for Policymakers[1]

The global average surface temperature has increased over the 20th century by about 0.6°C

- The global average surface temperature (the average of near surface air temperature over land, and sea surface temperature) has increased since 1861. Over the 20th century the increase has been 0.6 ±0.2°C. These numbers take into account various adjustments, including urban heat island effects. The record shows a great deal of variability; for example, most of the warming occurred during the 20th century, during two periods, 1910 to 1945 and 1976 to 2000.

Globally, it is very likely that the 1990s was the warmest decade and 1998 the warmest year in the instrumental record, since 1861

- New analyses of proxy data for the northern hemisphere indicate that the increase in temperature in the 20th century is likely* to have been the largest of any century during the past 1000 years. It is also likely that, in the northern hemisphere, the 1990s was the warmest decade and 1998 the warmest year.

---

* IPCC uses these words to indicate its judgement of the probabilities: very likely (90-99% chance); likely (66-90% chance).

• On average, between 1950 and 1993, night-time daily minimum air temperatures over land increased by about 0.2°C per decade. This is about twice the rate of increase in daytime daily maximum air temperatures (0.1°C per decade). This has lengthened the freeze-free season in many mid- and high- latitude regions. The increase in sea surface temperature

over this period is about half that of the mean land surface air temperature.

Temperatures have risen during the past four decades in the lowest eight kilometres of the atmosphere

• Since the late 1950s (the period of adequate observations from weather balloons), the overall global temperature increases in the lowest eight kilometres of the atmosphere and in surface temperature have been similar at 0.1°C per decade.

• Since the start of the satellite record in 1979, both satellite and weather balloon measurements show that the global average temperature of the lowest eight kilometres of the atmosphere has changed by +0.05 ±0.10°C per decade, but the global average surface temperature has increased significantly by +0.15 ±0.05°C per decade. The difference in the warming rates is statistically significant. This difference occurs primarily over the tropical and sub-tropical regions.

Snow cover and ice extent have decreased

- Satellite data show that there are very likely to have been decreases of about 10% in the extent of snow cover since the late 1960s, and ground-based observations show that there is very likely to have been a reduction of about two weeks in the annual duration of lake and river ice cover in the mid- and high latitudes of the northern hemisphere, over the 20th century.

- There has been a widespread retreat of mountain glaciers in non-polar regions during the 20th century.

- Northern hemisphere spring and summer sea-ice extent has decreased by about 10 to 15% since the 1950s. It is likely that there has been about a 40% decline in Arctic sea-ice thickness during late summer to early autumn in recent decades and a considerably slower decline in winter sea-ice thickness.

Global average sea level has risen and ocean heat content has increased

- Tide gauge data show that global average sea level rose between 0.1 and 0.2 metres during the 20th century.

- Global ocean heat content has increased since the late 1950s, the period for which adequate observations of sub-surface ocean temperatures have been available.

Figure 1, also taken from the IPCC's Summary for Policymakers, shows variations in the global surface temperature over the past 140 years.

Figure 1

Variations in the Earth's surface temperature over the last 140 years

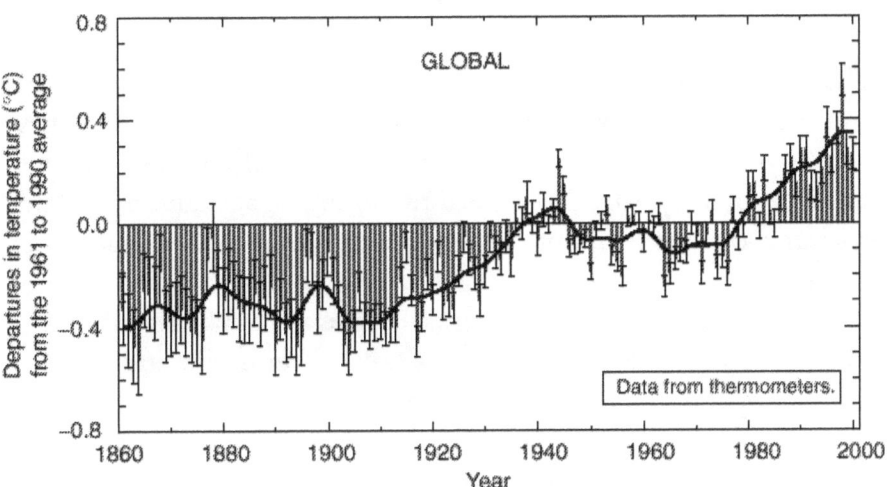

The Earth's surface temperature is shown year by year (red bars) and approximately decade by decade (black line, a filtered annual curve suppressing fluctuations below near decadal time-scales). There are uncertainties in the annual data (thin black whisker bars represent the 95% confidence range) due to data gaps, random instrumental errors and uncertainties, uncertainties in bias corrections in the ocean surface temperature data and also in adjustments for urbanisation over the land.

Over both the last 140 years and 100 years, the best estimate is that the global average surface temperature has increased by between 0.4 and 0.8 degrees C.[2]

The IPCC's findings appear to be reinforced by longer term data derived from measurements of tree rings, shallow ice cores, and corals, and from other methods of indirectly determining climate trends: these suggest that global surface temperatures are now as warm as or warmer than at any time in the past 600 years.

This evidence appears compelling at first sight, but a cautionary note is warranted. Many will remember that the climate was actually growing colder during the 1960s and early 1970s. There was widespread speculation that the world was heading for an ice age, and in Britain, a television programme about climate change called *The ice age cometh*, was screened in the early 1970s. The fact is that the range of normal climate variation is large, and genuine, long term changes in climate can be identified only after many years. As well, extreme climatic events are nothing new, and neither are new climate records.

What could be causing climate change?

Natural events cause changes in climate. For example, large *volcanic eruptions* put tiny particles in the atmosphere that block sunlight, resulting in a surface cooling of a few years' duration. One of the largest volcanic eruptions this century was that of Mount Pinatubo in the Philippines on 12 June 1991, which injected about 20 million tonnes of sulphur dioxide into the stratosphere together with enormous amounts of dust. This stratospheric dust caused spectacular sunsets around the world for many months following the eruption. The amount of radiation from the sun reaching the lower atmosphere fell by about 2 per cent. Global average temperatures fell by a bout a quarter of a degree Celsius for the next two years. There is also evidence that some of the unusual weather patterns of 1991 and 1992—for instance unusually cold winters in the Middle East and mild winters in Western Europe—were linked with effects of the volcanic dust.

Variations in *ocean currents* change the distribution of heat and precipitation. El Niño events (periodic warming of the central and eastern tropical Pacific Ocean) typically last one to two years and change weather patterns around the world, causing heavy rains in some places and droughts in others. Over longer time spans, tens or hundreds of thousands of years, natural changes in the geographical distribution of energy received from the sun and the amounts of greenhouse gases and dust in the atmosphere have caused the climate to shift from ice ages to relatively warmer periods, such as the one we are currently experiencing.

Some scientists have suggested that climate variations, even short-term ones, might be the result of changes in the *sun's energy output*. Such suggestions are bound to be speculative because no actual measurements of such changes exist. Accurate measurements of solar output have been available only since 1978, from satellites outside the disturbing effects of the Earth's atmosphere. These measurements indicate a very constant solar output, changing by about 0.1 per cent between a maximum and a minimum in the cycle of solar magnetic activity indicated by the number of sunspots. Astronomical records, and measurements of radioactive carbon in the atmosphere, show that this solar sunspot activity has sometimes, over the past few thousand years, shown large variations. Of particular interest is the period known as the Maunder Minimum in the 17th century when very few sunspots were recorded. Such a decline in solar output may have been a cause of the cooler period at the time known as the 'Little Ice Age'. Careful studies have estimated that the maximum variations in solar irradiance since 1850 are unlikely to be greater than about 0.5 watts per square metre. It is thought that this is about the same as the change in the energy regime at the Earth's surface due to about ten years' increase in greenhouse gases at the current rate.

Could changes in the *Earth's orbit* be contributing to climate change? Over the past 10 000 years, because of changes in the Earth's orbit, the solar radiation incident at 60°N in July has fallen by about 35 watts per square metre. But over 100 years the change is far smaller—only at most a few tenths of a watt per square metre. This is thought to be much less than the changes due to other factors, notably human activities. In any case the Earth's orbital changes have altered only the distribution of incoming solar energy over the Earth's surface; the total amount of energy reaching the Earth is hardly affected.

The Greenhouse Effect

Climate changes caused by human activities, most importantly the *burning of fossil fuels* (coal, oil, and natural gas) and deforestation, are thought to be far more important than the 'natural' changes in climate resulting from interactions such as those between the atmosphere and ocean, referred to as internal factors, and from external causes, such as variations in the sun's energy output and in the amount of material injected into the upper atmosphere by explosive volcanic eruptions.

Earth's atmosphere acts like a greenhouse, warming our planet in much the same way that an ordinary greenhouse warms the air inside its glass walls. Like glass, the gases in the atmosphere let in light yet prevent heat from escaping. This natural warming of the planet is called the greenhouse effect. The greenhouse effect that keeps our planet warm is really the "natural greenhouse effect". As the sun's energy reaches Earth's surface, some of it is reflected back and some absorbed. The absorbed energy warms the earth, which in turn radiates heat back towards space as infrared energy. Water vapour, carbon dioxide and other gasses in the atmosphere absorb some of the outgoing infrared energy, which heats them. These molecules then radiate the energy in all directions, including back to Earth. In effect, some of the energy remains trapped in our atmosphere, warming the planet.

Greenhouse gases—carbon dioxide, methane, nitrous oxide, and others—are transparent to certain wavelengths of the Sun's radiant energy, allowing them to penetrate deep into the atmosphere or all the way to Earth's surface. Clouds, ice caps, and particles in the air reflect about 30 percent of this radiation, but oceans and landmasses absorb the rest, and then release it back toward space as infrared radiation. The greenhouse gases and clouds effectively prevent some of the infrared radiation from escaping; they trap the heat near Earth's surface where it warms the lower atmosphere. If this natural barrier of atmospheric gases were not present, the heat would escape into space, and Earth's mean global temperatures could be as much as 33 degrees Celsius cooler (about -18 degrees Celsius as opposed to 15 degrees Celsius).

Over the centuries, the concentration of greenhouse gases, especially carbon dioxide, has fluctuated naturally, and the greenhouse effect has moderated the temperature of Earth accordingly. Now, our efforts to provide for Earth's growing population are releasing greenhouse gases into the atmosphere at rates greater than any other phenomena. As we burn fossil fuels, clear forests, and continue to use gasoline-dependent transportation, we increase the level of carbon dioxide and other gases in the atmosphere. As a result, we continue to harm Earth's atmosphere. The greenhouse effect that could be causing climate change is the 'enhanced greenhouse effect'. It works the same way as the natural greenhouse effect, but the extra carbon dioxide and other gases that we release into the atmosphere help increase the amount of energy that becomes trapped.

To understand how the natural greenhouse effect can impact the climate of a planet, we can look to our two heavenly neighbours, Venus and Mars. Venus, with a thick atmosphere of carbon dioxide and a "runaway greenhouse effect", has an average surface temperature of 480 degrees Celsius. Mars, with a thin atmosphere of carbon dioxide and virtually no greenhouse effect, has a mean surface temperature of minus 60 degrees Celsius. Earth seems to have an atmosphere and greenhouse effect that are just right for creating prime conditions for life.

The concern is that human activity is upsetting this balance, but any studies that try to identify human influences on climate have to separate a human-caused climate-change factor (the signal) from the background noise of natural climate variability.

Comparisons between observed patterns of temperature change and those predicted by models have now been made at the Earth's surface and in vertical sections through the atmosphere. Model predictions show increasing agreement with changes observed over the past 30-50 years. The closest agreement occurs when the combined effects of greenhouse gases and sulphate aerosol particles are considered. Statistical analyses have shown that these correspondences are highly unlikely to have occurred by chance.

But there are still uncertainties in these detection and attribution studies. These are due primarily to our imperfect knowledge of the true climate-change signal due to human activities, to our incomplete understanding of the background noise of natural climatic variability against which this signal must be detected, and to inadequacies in the observational record. Such uncertainties make it difficult to determine the exact size of the human contribution to climate change. Nevertheless, it would be hard to disagree with the cautious words of the IPCC's Second Assessment Report that "The balance of evidence suggests a discernible human influence on global climate".

The effects of climate change

The annual costs of climate change have been estimated at hundreds of billions of dollars. Such estimates are inevitably uncertain, and result from extrapolating the current cost of warming on agriculture, forestry, fisheries, energy, water supply, infrastructure, hurricane damage, drought damage, coast protection, land losses caused by rising sea levels, loss of wetlands, pollution, and so on.[3] Most ecosystems cannot respond or migrate as fast as the envisaged climate change. Natural ecosystems will therefore become increasingly unmatched to their environment. How much this matters will vary enormously from species to species: some are more vulnerable to changes in average climate or climate extremes than others. But all will become more prone to disease and attack by pests.

One effect of climate change is likely to be a *rising sea level*, arising from mainly from melt from glaciers and ice caps, the thermal expansion of the oceans, and to a lesser extent, changes in the Antarctic and Greenland ice-sheets. The IPCC's 'business as usual' scenario (no change in greenhouse gas emissions from trend) projects that the total average sea-level rise will be about 12 cm by 2030 and about 50 cm by the year 2100. Sea-level rise, though, will not be uniform over the globe. So while the average rises in sea level may not sound significant, many people living in coastal zones around the world will be affected. Many of the lowest lying coastal regions are fertile and densely populated. To people in these areas even a fraction of a metre increase in sea level can add enormously to their problems. Examples of especially vulnerable areas are the Pacific islands, the Netherlands and Bangladesh. About 6 million people in Bangladesh live on land that is less than one metre about sea level. It would be impractical to consider full scale protection of the long and complicated Bangladesh coastline from sea-level rise. So the most obvious effect will be that substantial amounts of good agricultural land will be lost—a serious problem for a country half of whose national income comes from agriculture, and for the very large proportion of the population that is at the very edge of subsistence.

But the loss of land would not be the only effect of sea-level rise. Bangladesh is extremely prone to damage from *storm surges*. Two recent surges, in November 1970 and April 1991, caused the losses of over 250 000, and 100 000 lives, respectively. Even small rises in sea level add to the vulnerability of the region to such storms. It is difficult to see what people in the affected agricultural areas will be able to do to relocate or adapt.

Global warming will tend to *exacerbate degradations due to human activities*. Sea-level rise will make the situation worse for low-lying land that is subsiding because of the withdrawal of groundwater and because the amount of sediment required to maintain the level of the land has been reduced. The loss of soil due to overuse of land or deforestation will be accelerated with increasing droughts or floods in some areas. In other places, extensive deforestation will lead to drier climates and less sustainable agriculture. Water supplies, which are in any case becoming increasingly critical in many places, are going to suffer in those parts of the world that are expected to become warmer and drier, especially in summer, with a greater likelihood of droughts; in other parts a greater incidence of floods is expected.

Because of adaptation to different crops and practices first indications are that total world food production will not be seriously affected by climate change. However, the disparity in per capita food supplies between the rich and poor countries will almost certainly widen.

There is expected to be a serious impact on *natural ecosystems*, especially at mid to high latitudes. Forests especially will be affected by increased climate stress causing substantial die-back and loss of production, associated with which is likely to be the positive feedback of additional carbon dioxide emissions. In a warmer world longer periods of heat stress will have an effect on human health; warmer temperatures will also encourage the spread of certain tropical diseases such as malaria to new areas.

Adaptation to climate change will often require changes in infrastructure, such as new sea defences or water supplies. But there may be benefits, and the uncertainties are daunting:
'It should be considered good fortune that we are living in a world of gradually increasing levels of atmospheric carbon dioxide. The satellite data on global temperature changes are now in. There has been no appreciable warming. ...Unlike other natural resources essential for food production which are costly and progressively in shorter supply, the rising level of atmospheric carbon dioxide is a universally free premium gaining in magnitude with time on which we call reckon for the future', writes Dr Sylvan Wittwer, in *The Great Promise of the Greenhouse Effect*.[4] Dr Wittwer is writing about increasing levels of carbon dioxide, but others believe that climate change *in itself* will be a good thing: leading to longer growing seasons and higher yields. They point to the boost to crop productivity given by higher levels of atmospheric carbon dioxide, which stimulate photosynthesis, enabling the plants to fix carbon at a higher rate. In glasshouses additional

carbon dioxide is sometimes introduced artificially to increase productivity. Under ideal conditions it can be a large effect: for doubled carbon dioxide, up to 25 per cent for wheat and rice and up to 40 per cent for soybeans. In September 2001 researchers using satellite data found that plant life above 40 degrees north latitude has been growing more vigorously since 1981 because, they believe, of rising temperatures and the build-up of greenhouse gases. 'The area of vegetation has not extended, but the existing vegetation has increased in density, marking an unexpected effect of global climate change.[5] For the world as a whole, though, the positive effects of warming and increased atmospheric carbon dioxide are likely to be more than outweighed by their negative effects in the same or other regions. Climate change certainly *has the potential* to inflict serious harm on many people. Estimates of the costs of climate change are obviously uncertain. But what *is* certain is that whatever happens to the climate, the developing countries have less capacity to adapt and so will suffer more.

Kyoto: the IPCC's response

The December 1997 Kyoto Protocol ('Kyoto') saw 159 nations reach the world's first legally binding commitments to reduce the global output of carbon dioxide and five other gases thought to contribute to the 'greenhouse' effect. Kyoto required industrialised countries to cut their emissions of the six greenhouse gases by an average of 5.2 per cent below their 1990 levels.[6] However, this target proved too ambitious, and the final agreement, reached in Bonn on 23 July 2001, committed the developed countries to smaller cuts, estimated to amount to about 2 per cent of their 1990 level. The United States, responsible for about 36 per cent of the world's greenhouse gas emissions, has refused to ratify the protocol, which will anyway not come into force until 55 countries, representing at least 55 per cent of 1990 carbon dioxide emissions, have ratified. (Signing a Protocol indicates a good faith *intention* to consider becoming legally bound by its terms, but it is ratification that is the formal act by which a country becomes bound under international law to comply fully with its obligations under the Protocol.)

The targets cover emissions of the six main greenhouse gases, namely:

- carbon dioxide (CO_2),
- methane (CH_4),
- nitrous oxide (N_2O),
- hydrofluorocarbons (HFCs),
- perfluorocarbons (PFCs)
- sulphur hexafluoride (SF_6).

One contentious part of the compromise is the freedom it gives to countries to meet some of their pollution reduction targets by using 'carbon sinks'—trees and other vegetation, which absorb carbon. This means they can make smaller and more electorally acceptable cuts in emissions from industry and transport. All changes in emissions, and in removals by sinks, go into the same basket for accounting purposes.

Kyoto also establishes three innovative mechanisms, known as joint implementation, emissions trading and the clean development mechanism, which are designed to help countries reduce the costs of meeting their emissions targets by achieving or acquiring reductions more cheaply in other countries than at home (see Box). While these mechanisms have been agreed in principle, their operational details need still to be fleshed out.

Three innovative mechanisms

Joint implementation

Joint implementation allows developed countries to share costs and credits for projects that reduce greenhouse gases or enhance sinks. This means that one country could do something that reduces carbon dioxide levels in another country, such as replanting a logged-out forest, or modernising a smoke-emitting smelter, and then apply part of that reduction against its own commitments.

Emissions trading

Emissions trading is a market system that allows those who own more emission units than they need to trade them to those who need more. So an organisation that owned 200 emission units, but emitted only 180 tonnes of carbon dioxide equivalent, would have 20 emission units to trade to a business that needed to use more emission units that it owned.

Domestic emissions trading refers to a regulatory regime in which specified businesses and other organisations would have obligations to report their emissions and to hold or purchase a corresponding number of emission units. Responsible parties who deforest land, or who harvest Kyoto forests, would also have obligations to acquire the necessary number of emission units (or sink credits) for the carbon released upon deforestation or harvesting.

International emissions trading, for which the Kyoto Protocol also provides: this is the transfer of assigned amount between parties to the Protocol, either between governments or between persons within these countries who have been authorised to trade.

Clean development mechanism

The clean development mechanism allows developed countries to ear emission credits for greenhouse gas emissions reduction projects and some sinks projects that are undertaken in developing countries.

The reduced Kyoto targets are far lower than what some environmentalists had hoped for, and what some countries, most notably the European Union, had been advocating. It was clear to the Kyoto negotiators that the treaty would only slow, but not stop, the build-up of carbon dioxide and other greenhouse gases in the atmosphere. (Carbon dioxide, which is given off by fossil fuel combustion, is thought to be by far the most important of the man-made greenhouse gases that form an insulating blanket around Earth.) But subsequent evaluations by leading scientists indicate that the environmental effects may be so small as to be almost unnoticeable in the near term.

Criticisms of Kyoto

Wrongly specified objective

> "The Parties ... shall, individually or jointly, ensure that
> their aggregate anthropogenic carbon dioxide equivalent

emissions of … greenhouse gases do not exceed their assigned amounts, calculated pursuant to their quantified emission limitation and reduction commitments … and in accordance with the provisions of this Article, with a view to reducing their overall emissions of such gases by at least 5 per cent below 1990 levels in the commitment period 2008 to 2012." Excerpt from Article 3 of the Kyoto Protocol to the Convention on Climate Change.

The first and most obvious criticism of Kyoto is that it is not an agreement about achieving a stable world climate. It is an agreement about reducing net emissions of greenhouse gases. Today's science suggests that there is a linkage, so that reducing net emissions of greenhouse gases will lead to a more stable climate, but there is no certainty about the relationships involved. Kyoto embodies 1990s science and the assumption that cutting greenhouse gases is the best way of achieving a more stable climate. It is, in this author's view, entirely unjustified to make these assumptions and then to impose massive costs on large numbers of people in support of an objective that may turn out to be irrelevant or even counter productive.

Our knowledge of climate change and its causes and effects is expanding rapidly. It may sound far-fetched now, but it is entirely possible that we shall find ways to reduce climate change drastically, and even to withdraw greenhouse gases from our atmosphere, far more efficiently than envisaged by Kyoto. But if that were to happen, Kyoto, with its extremely costly greenhouse gas emission commitments, will continue regardless, imposing huge costs on our economies.

The emphasis on complying with Kyoto could stifle research into investigating causes of climate change other than greenhouse gas emissions, and solutions other than reducing humanity's (net) emissions. Kyoto could overshadow these approaches, and divert research funds away from them.

Too ineffectual

In an analysis published in the journal *Science*[7] (January 1998), Bert Bolin, a Swedish meteorologist and the outgoing chairman of the United Nations Intergovernmental Panel on Climate Change, predicted that levels of carbon dioxide in the atmosphere will climb to 382 parts per million by 2010 if countries comply with their original Kyoto commitments. That would represent an increase of 8 per cent from 1990 levels, but a decline of only 0.4 per cent from the level it would have been had no actions were taken. Other scientists have estimated that the agreement originally envisaged by Kyoto would slow the projected rise in global temperatures by one-tenth to two-tenths of one degree Celsius by 2050.

Such a reduction would be 'an important first step' but would be 'far from what is required to reach the goal of stabilising the concentration of in the atmosphere,' wrote Bolin. Other climate experts agree with the essence of Bolin's conclusions, if not with all the specifics. A model by Tom Wigley, one of the main authors of the reports of the UN Climate Change Panel, shows how an expected temperature increase of 2.1°C in 2100 would be reduced by Kyoto to an increase of 1.9°C instead. Or, to put it another way, the temperature increase that the planet would have experienced in 2094 would be postponed just six years, to 2100.[8]

Exclusion of developing countries

Another major flaw is that while Kyoto provides for an emissions-trading scheme and other market-based mechanisms to compliance somewhat easier, it does not oblige developing countries to accept binding limits on their emissions in the near future. As emissions of greenhouse gases are rising fastest in the developing countries, worldwide emissions of greenhouse gases will continue to rise, even if the developed countries succeed in cutting back on their own emissions.

Too pessimistic

Some believe it likely that through new technology, stimulated by the price mechanism, alternative sources of energy will assume greater importance. Renewable energy – and especially solar power – may become competitive with, or even out-compete, fossil fuels by mid-century, according to some commentators.[9] This would make carbon emissions more likely to follow Kyoto's low emission projections, leading to global warming of about 2-2.5°C.

The IPCC itself says that global warming will not decrease *global* food production, that it will probably not increase storminess or the frequency of hurricanes, ["there is no general agreement yet among models concerning future changes in mid latitude storms (intensity and frequency) and variability," and that "there is some evidence that shows only small changes in the frequency of tropical cyclones." Nor will it increase the impact of malaria.[10]

Extremely expensive

The cost of dealing with climate change is huge. For the US alone the cost of complying with Kyoto would be higher than the cost of solving the single most pressing problem for the world—providing the entire world with clean drinking water and sanitation. It is estimated that the latter would avoid 2 million deaths *every year* and prevent half a billion people from becoming seriously ill. If it imposes controls on the industrial world the cost of Kyoto would approach $1000 billion, or almost five times the cost of worldwide water and sanitation coverage. Compare this with today's total aid budget of around $50 billion annually.[11]

Weaknesses of trading mechanisms

Most businesses and environmentalists support the general idea of emissions trading for all its faults. Whether governments and companies get credit for CO_2 reduction projects may depend on any number of factors—from the variety of trees being replanted to local legislation on fuel usage. But because Kyoto reductions targets are set against 1990 emissions levels, economically troubled countries like Russia and Ukraine now have a huge number of credits to sell—not because they've become more energy efficient, but because their lowered industrial output has led to lower emissions. Environmentalists call these credits "hot air," and say trading them won't help the earth. Other environmentalists, while agreeing that trees do indeed absorb carbon emissions and release oxygen, so removing carbon dioxide from the atmosphere, believe the developed countries should not be able to get away with simply planting trees as an alternative to cutting emissions: the Kyoto accord does indeed limit how many credits can be earned through afforestation projects. There are other details to be worked. Researchers also are trying to come up with harder data on how to apportion credits for CO_2 absorbed by trees—and there is also the question of what do with these credits if the forest burns down.

Political fallout

Some fear that governments will cite the modest environmental benefits of complying with their Kyoto obligations as an excuse for doing nothing else. Even if all industrialised countries honour their commitments to reduce atmospheric pollution, levels of heat-trapping greenhouse gases in the atmosphere will continue to grow.

Discussion

A brief summary of climate change and Kyoto would make these points:

- It is highly likely, but not proven, that the climate is changing and that human activities, including specifically those that release greenhouse gases into the atmosphere, are causing these changes.

- Climate change is likely to be causing some weather related events, and to cause even more in the future. Most of these events and effects are negative for human, plant and animal life, but some are likely to be positive.

- Kyoto addresses greenhouse gases solely; it will certainly be costly, both financially and politically, and may well be ineffectual. Some suggest it would be cheaper to adapt to climate change, rather than try to stop it.

- Kyoto will be immensely costly and will divert resources away from other possible solutions to the negative effects of climate change, as well as from other global social and environmental problems.

Such arguments may seem petty compared to the potentially disastrous effects of climate change, and the likelihood that the most grievous effects will fall on those least able to cope. Despite the very significant scientific uncertainties and reservations about Kyoto itself, it would be irresponsible to do nothing about climate change. The changes in climate that we already appear to have experienced, and that we may be destined to experience in the near future, should not be ignored only because there is no definitive proof that it is happening, or that cutting back on greenhouse gas emissions may not be the optimal solution.

Kyoto suffers from the same conceit as many other government approaches to social and environmental problems. It embodies the assumption that government knows the best way of achieving its goals, and that this way is to control humanity's net greenhouse gas emissions. It would seem, though, poor policy to impose extremely costly controls on human activities, on the basis that they *might* help bring about a stable climate.

An ideal way of addressing climate change would be to find a mechanism that does not have to embody the assumption that it knows exactly how the Earth's climate is changing, what is causing it to change, and what is the best way of solving dealing with any change. It would encourage innovative solutions and it would not assume a degree of scientific certainty that is just not there. It would be as cost-effective as possible, as the cost of dealing with global climate change is bound to be very high. It would stimulate the investigation and adoption of promising new technologies, and be open to new information about the causes and effects of climate change. It would most probably seek to constrain the negative effects of climate change, while doing little to discourage positive effects.

Ideally too, it would use markets, the best way yet devised of allocating society's scarce resources, to channel people's self-interest into the solution of the climate change problem.

If such a solution could be found, it would be bound to attract more support from world leaders, non-governmental organisations, and the public in general than Kyoto. Given that any solution is going to involve large costs and sacrifices, such support is essential.

Chapter 2

Better than Kyoto: Climate Stability Bonds

This chapter is in three sections. Section A outlines the concept of Climate Stability Bonds, a new financial instrument designed to achieve climate stability, rather than to regulate activities or institutions. Section B answers some questions about how the Bonds would work. The final section, C, quickly summarises some of the advantages that Climate Stability Bonds offer over Kyoto. These and other advantages are elaborated in the next chapter.

A: Climate Stability Bonds

Climate Stability Bonds are a new, and as yet untried, financial instrument. A fixed number of Climate Stability Bonds would be issued, and auctioned to the highest bidders. The issuers would undertake to redeem the Bonds for a fixed sum *only when the climate has achieved a targeted level of stability*. Normal bonds are redeemable at a fixed date, for a fixed sum, and so yield a fixed rate of interest. Climate Stability Bonds are entirely different: they would not bear interest and their redemption date would be uncertain. Bondholders would gain most by ensuring that climate stability is achieved quickly. In this way there is no need for the issuing body to make assumptions as to *how* to stabilise the world climate—that is left to Bondholders.

The issuing body could be an agent of, or supervised by, an international organisation such as the United Nations or World Bank. It would undertake to redeem the Bonds using funds that could perhaps be obtained from all countries, in proportion to their Gross National Product. It would be up to individual countries to decide how to raise funds, which they would probably do from taxation revenue.

Internationally backed Climate Stability Bonds would be issued by open tender, as at an auction; only those who bid the highest price for the limited number of Bonds would be successful in buying them. A fixed number of Bonds would be issued, each redeemable for, say, $1 million, only when climate stability, as defined by the issuers and as certified by objective measurement, has been achieved and sustained. Once issued, the Bonds will be freely tradable on the free market at any time until redemption.

What will determine the price of the Bonds? Most obviously, the market's assessment of how close climate stability is to being achieved. Interest rates on alternative investments will also be a factor. The Bonds would sell for a small fraction of their redemption value if people thought there were virtually no chance of climate stability being achieved in their lifetime. People will differ in their valuation of the Bonds, and their views will change over time. But the Bonds, once issued, would be tradable on the free market at any time. Bondholders, having done their bit to achieve climate stability, would want to realise the capital gain arising from the higher market price of their Bonds. These market prices would be publicly quoted, just like those of ordinary bonds or shares.

The way Climate Stability Bonds generate incentives is simple: assume that Climate Stability Bonds, each redeemable for $1 million have been issued, and that when floated they each sell for $100 000. People or, more likely, institutions now hold an asset that can give them a return of 900 percent once a stable climate has been achieved. They may not be able to do very much individually, but they have a powerful motivation to *cooperate with other bondholders* to bring about a more stable climate.

There are obvious difficulties involved in defining what a stable climate actually is, *but the same difficulties apply when attempting to monitor the success or otherwise of Kyoto.* How are we to know whether Kyoto's expensive cuts in greenhouse gas emissions are taking us closer to our objective of a more stable global climate? Presumably scientists will measure the effects of the cuts by monitoring such objectively verifiable indicators as temperature, change in temperature, rate of change of temperature, precipitation, frequency of extreme climatic events, and many others, at a wide range of locations.

Under a Climate Stability Bond regime, the Bonds would be redeemed only when climate stability, as defined by a similar set of indicators, had been achieved. A Bond regime might also explicitly target less scientific measures, such as the frequency and severity of adverse climatic events, the numbers of people killed or made homeless by such events, or the insurance payouts to which they give rise.

Climate Stability Bonds, once issued and sold, must be readily tradable at any time until redemption. This is critical to the operation of the Climate Stability Bond mechanism. Many bondholders would want, or need to sell their Bonds before redemption—which may be a long time in the future. If there were no possibility of selling their Bonds they would not be able to realise any capital appreciation, which would remove much of the incentive to purchase the Climate Stability Bonds when issued.

But there is another important reason for requiring a healthy market in Bonds after issue: active investors may be able to speed up only one, or a few, of the processes necessary to stabilise the world's climate. Once these investors have done what they can, and seen the capital value of their Bonds increase in line with the resulting increased probability of the Bonds' early redemption, they may have no wish to speculate on the speed at which the remaining processes will be carried out. Other groups of active investors, who will have greater expertise in performing these later processes, must be given an incentive to use their skills to accelerate attainment of climate stability. The capital appreciation of Bonds bought from previous owners, and sold at a still higher price (or redeemed) provides that incentive.

Kyoto formalises the assumption that controlling the targeted greenhouse gases is the best way of achieving climate stability. Apart from the daunting uncertainties about the contribution of greenhouse gases in climate change, there is even less understanding of the role that agriculture and forestry can play as sinks for greenhouse gases.

In contrast, Climate Stability Bonds targeting climate stability would both bypass these, and other, uncertainties, and encourage research into discovering and clarifying the relevant scientific relationships. There is no need for the targeting mechanism to make assumptions as to how best to stabilise the world climate. Bondholders—and potential bondholders—would do this, not only as the Bonds are issued, but continuously, over the entire lifetime of a Climate Stability Bond issue, in the light of our ever-increasing scientific knowledge.

Obviously people will differ in their valuation of the Bonds. Their views will also change as events occur that make achievement of a stable climate a more or less remote prospect and as new information about climate, and about the causes of climate change, is discovered. So the market price of Climate Stability Bonds would be constantly changing, as are the prices of other financial instruments such as equities, currencies or conventional Bonds. The prices of Climate Stability Bonds, and the way they change with time and events, will convey information of the utmost value to policymakers.

What would Bondholders do?

A Climate Stability Bond regime would not dictate how to achieve a stable climate. Bondholders could undertake a wide range of projects such as:

- helping countries or companies to set up greenhouse gas emission control programmes or carbon sequestration plantations;

- financing the production or consumption of renewable energy resources, which would also reduce greenhouse gas emissions;

- carrying out, or supporting, research into increasing the Earth's albedo.

But much would depend on the precise definition of the targeted objective. If 'the number of people killed or made homeless by climate-related adverse events' were targeted for reduction as one component of a climate stability objective, then bondholders would probably help finance the construction of defences against floods or a rising sea level.

Bondholders might also find it worthwhile to lobby governments the world over to enforce existing pollution control policies, or to legislate in favour of new climate stabilising measures. As well, they can be expected to finance other research and initiatives, all aimed at stabilising climate as cost-effectively as possible, whose precise nature we cannot yet envisage

Some governments, research institutes and others are already carrying out these, or similar, activities. The crucial difference is that, under a Climate Stability Bond regime, the Bondholders will have the incentive to allocate funding amongst widely differing approaches efficiently. They will be strongly motivated to seek out those ways of achieving a stable climate that will give them the best return on what is ultimately the taxpayers' outlay. Only when the targeted degree of climate stability is achieved will governments have to pay for it by redeeming the Bonds. Until then, it is bondholders who have to finance the initiatives that they think will achieve climate stability. The body that issues the Bonds will, in effect, have contracted out the achievement of climate stability to the private sector – having defined the degree of climate stability it wants and undertaken to pay Bondholders when it has been achieved.

B: Operation

This section poses, and attempts to answer, questions concerning some aspects of the operation of the Climate Stability Bond mechanism.

How many Bonds would be issued?

The redemption value of each Climate Stability Bond, and the number of Bonds issued, would determine the *maximum* cost of achieving the targeted objective. Before issuing Climate Stability Bonds, the issuing body, in consultation with governments, would have to decide, very approximately, on this maximum value. This outcome need not be the desired ultimate level of climate stability: it could be some intermediate level of stability, greater than the current level, but less than some ultimate target. Note also that separate Bond issues could target different, complementary, climate stabilising objectives, either in sequence, or simultaneously.

A major consideration would be the financial benefit of the targeted increase in stability. For many countries' economies and governments, an increase in climate stability would bring about financial savings. With a more stable climate, expenditure on certain infrastructure projects could be reduced, as could contingency funds for weather-related disasters. But there would also be extremely significant non-financial benefits arising from a more stable climate. Governments, acting collectively, would take into account both types of benefit in deciding on the maximum value of the targeted increase in climate stability before issuing the Bonds. If Climate Stability Bonds are to be used in conjunction with other policy instruments aimed at achieving the same goal, the issuing body, in consultation with governments, would also have to decide on what proportion of the total expenditure would be spent on redeeming the Bonds.

These factors would determine how many Bonds the issuing body would float. The maximum cost of the Bond issue would equal: the total number of Bonds multiplied by each Bond's redemption value *plus* administration costs *minus* any revenues gained on floating the Bonds.

But while it has to decide on the maximum price it is prepared to pay to achieve the targeted rise in climate stability, the issuing body need not work out how much the *actual* cost will be with any accuracy: that would, in effect, be done by the bidders for the Bonds in competition with each other. Let us assume that Climate Stability Bonds are issued with the aim of raising the world's climate stability, and that the issuing body issues 5000 Bonds redeemable for $1 million each, once the objective has been achieved—this corresponds to Scenario 1 in Table 1(page 99). This means the maximum cost of achieving this objective, which would be paid if the Bonds fetched virtually nothing on flotation, will be 5000 multiplied by $1 million, which equals $5 billion. Assume that the price at which the Bonds are initially auctioned off, the float price, is just $100 000 each. This means that the market is judging that the net cost of achieving the specified climate stability objective will be $4.5 billion. Importantly, *this judgement will be independent of the number of Bonds issued.* To see how this affects the float price, assume now that the issuers are prepared to spend as much as $10 billion to achieve the same climate stability objective—this corresponds to Scenario 2 in Table 1. Instead of issuing 5000 Bonds they issue 10 000, each redeemable, again, for $1 million. They would then be liable for a maximum cost of $10 billion. Despite the higher potential total sum on offer, competition for the Bonds would mean that investors would still reckon on achieving the targeted level of climate stability for the same sum: $4.5 billion. With 10 000 Bonds on offer, this implies a difference between the redemption value and float price of $450 000, and so a float price of $550 000 per Bond. The effect of issuing a larger number of Bonds to achieve the same objective is simply to raise the float price of each Bond. (Climate Stability Bonds would be an unusual financial instrument, in that the more that are issued, the higher would be their value!) Scenario 3, under which 50 000 Bonds are issues, sees the float price of the Bonds rise to $910 000 each. The important point is that the issuing body does not have to do all the work in deciding how much the objective of climate stability will cost to achieve. Competitive bidding for the Bonds at issue will minimise the total cost of achieving the targeted level of climate stability, while the issuers can put a cap on its total liability by limiting the number of Bonds issued.

Under Kyoto, as in similar global or national undertakings, many of the people charged with achieving climate stability will be bureaucracies, government agencies, or large private sector operations. If they do not face effective competition in bids to undertake projects controlling net greenhouse gas emissions, they will have every incentive to inflate their costs. Even if they do have to compete to obtain contracts, they will have no continuing, powerful incentive to contain their costs. In contrast, the Climate Stability Bond mechanism ensures that competition and the market decide roughly how much climate stability will cost to achieve. Competition and market forces will operate when would-be investors bid for the Bonds at issue and *continue to do so* until the targeted increase in climate stability has been achieved and the Bonds redeemed. This fact, and the incentive it gives for bondholders to continuously seek to minimise their costs, contrast with Kyoto, in which the cost of achieving the desired outcome, if it is estimated at all, is not known, nor will it be subject to competitive bidding during the entire, long, time it will take to achieve its goal.

Note that the issuing body could add to the number of Bonds in circulation after floating, at any time, if it wanted to boost the efforts going into climate stabilising projects. It might do this, for example, if new research showed that the consequences of climate change on the environment would be worse than previously envisaged. Unlike shareholders in public companies, bondholders would have no objection to more Bonds being issued: rather than to dilute the value of their holdings, the effect would be to *increase* the market price of the Bonds they currently own.

If the issuers wanted, for whatever reason, to *reduce* the efforts going into achieving climate stabilising activities, the situation is a bit more complicated. They could buy Bonds back from holders, but doing so would reduce the total funds to be spent on achieving the targeted objective, and so lower the value of all Bonds in circulation. People may therefore be unwilling to buy Bonds in the first place if they thought there were a high probability of the issuers' buying them back in this way. They would demand some sort of premium for taking that risk. Or the issuers could undertake either that it would never buy Climate Stability Bonds back or, that if they did, they would undertake to buy them at the market value ruling before it announced its purchases.

Who would buy the Bonds?

Some people might purchase Climate Stability Bonds with the idea of doing nothing but holding on to them until they could sell them at a profit, especially if their price fell to very low levels. These *passive investors* would have no intention of doing anything to help achieve climate stability. Some of them would be casual purchasers, who would buy Bonds (or shares in a Bond) with the same intent as they would a lottery ticket or shares in companies that do not pay dividends. They would hope to hold on to the Bonds until redemption, or until their market value had risen sufficiently high for them to enjoy a worthwhile capital gain. Other passive investors would be speculators, who know, or think they know, that the climate will stabilise more quickly than the rest of the market thinks it will, and that the Bonds are therefore underpriced.

Another category of passive investor might be the hedger. These would be people who, if they were unable to buy Climate Stability Bonds, stand to lose if the climate becomes more stable. Hedgers might buy the Bonds as a form of insurance against this possibility. They might, for example, be global businesses involved in constructing sea defences or hurricane-proof buildings, whose long term prospects would be worsened by a fall in the number of adverse climate-related storms. Of course, if one of the indicators explicitly targeted by the Bonds were 'number of people killed by storms' (or similar) such businesses could benefit from increased climate stability.

Casual purchasers and speculators would want to become 'free-riders', hoping to benefit from any increase in the Bond price without actually participating in any climate stabilising projects. Hedgers wouldn't particularly want the value of their Bonds to rise, but their bondholding would similarly reduce the supply of Bonds available to active investors. However, the way markets work would limit the opportunities for all these passive investors. The more Bonds they collectively own, the more remote would the targeted objective becomes, the lower the market price of their Bonds would fall, and the more they would stand to lose as the aggregate value of their Bond holdings also falls (see box: *the Free Rider question*). At some point, then, it would become worthwhile for passive investors either to become, or to sell their Bonds to, *active investors*.

These people, or institutions, would use their own capital, or borrow on the strength of the redemption value of their Bonds, to initiate or facilitate climate stabilising programmes. Active bondholders would have a powerful incentive to cooperate with each other to launch these programmes, and to ensure that they are as cost-effective as possible. These people's motivation will come from the expected capital gain they will enjoy as the Bond price rises with the enhanced probability that the targeted increase in climate stability will be achieved quickly.

The Free Rider question

Could 'free riders' undermine the workings of the Climate Stability Bond mechanism? If purchasers of a significant number of the Bonds hold them with no intention of doing anything to increase climate stability, that would surely undermine the potential of a Bond regime.

There are, though, grounds to believe that free riding would not be a serious problem, mainly because it is unlikely much free riding would occur, and partly because even if it did occur, it would not impede the operation of the Bond mechanism. Free riding would be a self-cancelling activity: if most of the Bonds were held by would-be free riders, then little would be done to help bring about a more stable climate. And as the targeted climate stability objective becomes more remote, the value of the Bonds would fall. Because the Bonds' price would fall, they would make a more attractive purchase for those who would be prepared to help stabilise the climate. So would-be free riders would be tempted to sell, even at a loss, rather than see the value of their Bonds continue to fall. And an early history of falling Bond prices would tend to make free riding on later Climate Stability Bond issues less appealing.

There are other reasons why passive bondholding would be unattractive to potential free riders:

- Individual free riders would have no incentive to collude with other free riders, because the more they did so, the more remote the targeted objective would become, and the further would the value of their Bonds fall. This would act so as to restrict any free riding activity to small players.

- As with other financial instruments, small players would have to pay higher transaction costs than the bigger institutions – those who can initiate effective climate stabilising projects.

- Small players also would not have access to the research that would enable big players to value the Bonds accurately. Therefore they would be at a disadvantage in the market.

Note that even if free riders were to gain from holding Climate Stability Bonds, they would do so only if their Bonds rose in value, which would happen only if the targeted objective comes closer to being achieved. Note also that attempted free riding would have some positive effects: it would, for example, add liquidity to the Climate Stability Bond market.

What should Climate Stability Bonds target?

For Climate Stability Bonds to be effective, the targeted objective, and its associated sub-objectives and indicators, must be carefully defined, so that their movements either actually are, or are strongly correlated with, what society wants to achieve, which we will presume is a more stable climate. Unlike Kyoto, which targets (net) greenhouse gas emissions, Climate Stability Bonds allow a range of quite different indicators to be targeted.

What are the key criteria of sub-objectives that would make them suitable for inclusion as elements of an overall Climate Stability Bonds objective?

- They will have to be *carefully defined,* so that their achievement is, or is highly correlated with, what society, collectively, wants a more stable climate to be.

- They should be capable of being *accurately targeted by quantifiable indicators*, whose progress inevitably corresponds with progress toward the desired outcome.

Kyoto fails on the first criterion. It does not have a stable climate as an explicit objective, but only the cutting back of (net) greenhouse gases emitted by the developed countries. It does not even target the composition of the atmosphere, so it effectively devalues efforts to remove greenhouse gases already in the atmosphere. Nor does it allow for, admittedly unlikely, natural events that could alter atmospheric composition. Even in its own terms, then, its focus is too narrow. There is nothing in Kyoto that will link its perceived success or failure to (probably) desirable changes in the composition of the atmosphere, let alone to climate stability.

The virtue of targeting outcomes, as the Bond principle does, is that it would make it impossible for policymakers to avoid being clear about exactly what they want to achieve. Under a Climate Stability Bond regime, policymakers would have explicitly to answer the crucial question: what is the desired outcome?

Climate stability, for targeting purposes, could be defined as a function of many variables, which *need not be limited to those describing the climate.* Climate Stability Bonds could target a wide range of variables weighted according to their expected *effects* on humans and the wider environment. These variables can be quite diverse and could include some or all of, for example:

- Temperature, rate of change of temperature, rate of change of change of temperature, measured at sites around the world.

- Biodiversity, at selected sites, with special targeting of the well-being of 'indicator' species. The issuing body will need to remember that loss biodiversity can result from many causes, of which climate instability is only one. Climate Stability Bonds should use biodiversity as an indicator or objective only insofar as it is affected by climate.

- Depth of ice sheets.

- Direct effects of weather-related events on human life. Numbers of lives lost, or the insurance payouts consequent on these events, the expenditure on preparations against future events. Some index of these indicators could be calculated, with extra weighting if the events are thought to be independent of each other, and widely distributed around the globe.

In principle the targeted objective should be as broad as possible: this applies to any policy aimed at stabilising climate, not only to Climate Stability Bonds. To see why, let us look at Kyoto, which has chosen a very narrow objective: to reduce the net emissions of those gases thought (in the 1990s) to be responsible for the greenhouse effect. But there may be other factors that influence the climate: there may even be other greenhouse gases, the targeting of which would be more cost-efficient than those identified by Kyoto. A policy such as Kyoto that is focused solely on a narrow range of greenhouse gases may encourage people to increase their emissions of other gases, or to respond in other ways that defeat the object of the policy. Nor would Kyoto do anything to prevent the uptake of new technologies that may increase climate instability without increasing greenhouse gas emissions.

A Climate Stability Bond regime, whose sole objective was to reduce greenhouse gas emissions would, of course, have the same deficiencies. It is far better to target the desired outcome, which may have little to do with gas emissions, than the supposed means of achieving that outcome. *In principle ends, rather than means to ends, make better targets for policy.* Climate Stability Bonds can be tailored so as to target any desired range of effects of climate instability. Because they target outcomes, they will inevitably entail early clarification of what is actually wanted. We saw in chapter 1 that climate change may lead to some positive effects: it could, for example, lead to longer growing seasons and higher yields in some regions. Others point to the boost to crop productivity given by higher levels of atmospheric carbon dioxide. 'Climate Stability' as targeted by Climate Stability Bonds, could be defined such that Bondholders *tackle only the negative effects of climate change.*

How important is tradability?

Climate Stability Bonds, once floated, must be readily tradable at any time until redemption. The operation of such a 'secondary market' is critical to the way the Bonds work. Many Bond purchasers will want, or need, to sell their Bonds before redemption—which may be a long time in the future. With a secondary market, these holders will be able to realise any capital appreciation experienced by their holdings of Climate Stability Bonds, whenever they choose to do so. Tradability would make the Bonds a more attractive investment in the first place.

As the Bonds are traded, they would tend to flow towards those who are most able to achieve the targeted increase in climate stability. In fact, though, it is not necessary for there to be any actual flow of Bonds. Large bondholders might simply decide to subcontract out the required work to many different agents, while they themselves hold the Bonds from issue to redemption. The important point is that the Bond mechanism ensures that the people who allocate the finance have a continuing incentive to do so efficiently and to reward successful outcomes, rather than merely to pay people for undertaking an activity. At the limit we can conceive of just one single buyer of all the Bonds. If this buyer were determined to hold on to the Bonds until redemption, then the Bonds would function as a sort of performance-related contract, with the issuer paying out only when the climate has been stabilised. The buyer could contract out most, or all, of the work required to achieve the objective, with the incentives given by the Bonds for speedy accomplishment cascading down from the bondholder to those subcontracted to do the work.

Too large a number of small bondholders could probably do little to help solve the climate change problem by themselves. If there were many small holders, it is likely that the value of their Bonds would fall until there were aggregation of holdings by people or institutions large enough to initiate effective problem-solving projects. As with shares in newly privatised companies the world over, Bonds would mainly end up in the hands of large holders—individuals or institutions. Between them, these large holders would probably account for the majority of Bond holding. Even these bodies might not be big enough, on their own, to achieve much without the co-operation of other bondholders. They might also resist initiating projects until they were assured that other holders would not be free riders. So there would be a powerful incentive for all bondholders to *co-operate with each other* to help achieve climate stability. They would share the same interest in seeing this objective achieved quickly. So they would share information, trade Bonds with each other and collaborate on objective-achieving projects. They would also set up payment systems to ensure that people, bondholders or not, bring about all the objectives necessary to achieve climate stability. Bondholders would either trade Bonds, or make incentive payments to ensure that any proceeds from higher Bond prices, or from redemption, would be channelled in ways most likely to stimulate speedy achievement of climate stability. Large bondholders, in co-operation with each other, would be able to set up such systems cost-effectively.

So, regardless of who actually owns the Bonds, aggregation of holdings and the co-operation of large bondholders would ensure that those who help achieve climate stability are rewarded in ways that *maximise the increase in the world's climate stability per unit outlay.*

What about countervailing incentives?

There are two possibilities here.

1. Futures and options markets in Climate Stability Bonds could develop, enabling people to benefit from a falling Bond price, so giving them an incentive to delay climate stability.

It is quite likely that there would be futures and options markets for the Bonds, and it is almost certain that the price of any particular Climate Stability Bond would not always be increasing along an upward trend from its float price to its redemption value. It is right too that bondholders should be able to hedge against the consequent falls in the value of their asset, and it follows that people who do not hold Bonds will also participate in markets for derivatives of Bonds, some of which would rise in value as the climate became less stable. This in turn means that speculators and short sellers could certainly profit by *short-term* Bond price falls, and the question is whether these people would then take steps to reduce climate stability. There are two main reasons why they would not. The first is that, in the long term, the weight of money would be against them. Provided sufficient funds are allocated to achieving the climate stability objective, there would be a very large net positive sum of money payable if the climate is stabilised, and a net zero sum being paid if the climate is not stabilised. All the long-term incentive is to achieve a stable climate. Those who, for whatever reason, would suffer from an increasingly stable climate, could be compensated by bondholders, or bribed to change their ideas. Note also that for every buyer of a 'put' option there would be a seller, and that for every futures contract bought on the expectation that the Bond price would fall, there would be an equivalent futures contract sold on that basis, so that the net incentive generated by derivatives would be in line with the incentive created by the underlying financial instrument, the Climate Stability Bond: in the long run, this would be strongly in favour of increasing climate stability.

The other reason that short sellers, or holders of put options, in Climate Stability Bonds would not take actions aimed at destabilising the climate is that many such actions may well already be illegal or, again given the incentives that the Bonds would generate, be made illegal once the Bonds have been issued.

2.   Participating governments could try to avoid expenditure by failing to redeem the Bonds, either by reneging on their commitment, or by delaying climate stability.

The issuers of Climate Stability Bonds would represent participating governments, which would in turn be representing their population. They would therefore be under pressure, either from their own population, or from other governments, to comply with their commitment to supply funds for Bond redemption, and not to destabilise the climate. But it is also in governments' own interest to fulfil their obligation. If they did not, they would be discrediting the whole Bond principle, which they may well want to use domestically, or to secure further increases in global climate stability, or as participants in other global efforts to solve a different social or environmental problem. If the issuing body still thought that governments would renege on their commitment, they could insist on holding funds for redemption in escrow.

What happens once the objective has been achieved, and the Bonds redeemed?

Once a more stable climate is close to being achieved, the issuing body can float a new set of Climate Stability Bonds aimed at maintaining the achieved outcome, or at further improvements in the climate's stability. The objective, of course, is a *sustained* improvement in climate stability, and this is how it would have to be defined when the Bonds are issued. Sustaining the outcome beyond the period specified in the original Bond issue is likely to be cheaper than achieving it, while further improvements in climate stability targeted by subsequent Bond issues are likely to cost less, in terms of benefit per unit outlay, than those achieved by the first issue. There are three main reasons for this, the first two of which are linked:

1. Bondholders may have invested in systems or capital assets that cost less, per unit benefit, to keep running than they did to set up.

2. Bondholders, in a similar fashion, would have learned from their experience of achieving the objective targeted by the first

Bond issue. They would have looked for, and experimented with, different methods of increasing climate stability, and be able to choose the most efficient ones for subsequent Bond issues. Any know-how about, for example, monitoring systems or equipment installation, or how to persuade recalcitrant governments to enact and enforce appropriate legislation would be more cheaply available once climate stabilising projects are up and running.

3.   Less specifically, it is likely that general improvements in productivity, mainly arising from technology (including information technology), will continue to occur in our economies, and that bondholders will continue to adopt and adapt them.

C: Advantages of Climate Stability Bonds

Climate Stability Bonds have two critical, linked, advantages over Kyoto. The first is that the Bonds do not rely on the robustness of our existing scientific knowledge. And the scientific uncertainties are legion, involving, for example, questions about:

•   whether climate change is happening,

•   the degree to which climate change may be happening,

•   the implications of climate change for plants, animals and humans in different parts of the world, and

•   the causes of climate change, and the way these causes interact.

There are also huge financial uncertainties about the costs of attenuating or eliminating climate change. Kyoto aims to reduce emissions of a small range of gases. But, as discussed in chapter 1, there may be other causes of climate change that are far more important, of which we are currently unaware. A Climate Stability Bond regime, targeting climate change directly, may well lead to cuts in greenhouse gas emissions, but it would *not assume that doing so is the best solution.* Climate Stability Bonds improve on Kyoto, because they encourage behaviour leading to the desired outcome, rather than seeking to control activities whose effects on climate stability are not fully known.

For all these reasons a Climate Stability Bond regime would have obvious advantages over Kyoto. Bondholders will probably find it at least as worthwhile to continue to investigate the causes of climate change than to cut back on emissions of greenhouse gases, as envisaged by the Kyoto policymakers. Their research might even lead them to conclude that climate change is a result of a combination of circumstances that will resolve itself without any need for intervention. Whatever bondholders do, they will be motivated to achieve a stable climate using the most cost-effective combination of measures possible.

For climate change, as with most environmental problems, not all the alternative solutions are known in advance. The optimal range of approaches is seldom a one-size fits all, government-dictated, inflexible, policy suite. More often, it is a matter for investigation and experimentation, and a wide variety of approaches is essential. Under a Climate Stability Bond regime, diverse activities would be rewarded in proportion to their success. And once applied they, or an optimal combination of them, would become the new standard for future projects.

The other major advantage of a Climate Stability Bond regime is that bondholders tackle whichever causes of climate change give them the best return for their outlay. Incentives to be efficient are built into every process necessary to achieve a stable climate. Competitive bidding for the Bonds at issue, and at any time until redemption, minimises the costs to taxpayers. The more efficient that bondholders are in achieving climate stability the more they will gain from appreciation of the value of their Bonds. This efficiency maximises the degree of climate stability that can be achieved per dollar outlay. Because of the colossal sums involved, the absolute value of the benefits that Climate Stability Bonds offer in comparison to Kyoto are likely to be huge.

Further advantages of a Climate Stability Bond regime are:

- the informational advantages that apply because there are (probably) large numbers of contributors to climate change. Kyoto, in effect, seeks to control many sources of pollution, as greenhouse gases are emitted from many sources. About half of carbon dioxide emissions, for instance, come from dispersed sources, such as cars and home heating systems. Regulation of the sort that a Kyoto regime will encourage can work efficiently when dealing with intrinsically large-scale processes. But when it comes to emissions of greenhouse gases, there are very large numbers of polluters, which can make enforcement of environmental regulations expensive and intrusive. Establishing and monitoring a fully comprehensive carbon dioxide emission regime, for example, is going to require immense quantities of information. Climate Stability Bonds, focused as they are on the desired outcome rather than activities, would target society's overall objective – climate stability– rather than each of the numerous activities that (supposedly) generate climate change. They would therefore require comparatively few data, and less intrusive monitoring: regular assessments of the climate need only be taken at selected sites.

• that funds for climate stability need not be used for scientifically approved projects. They could, for instance, be used to bribe corrupt or malicious governments so as to encourage them to look favourably at climate stabilising legislation or projects in their countries. Appealing to these governments' financial self-interest could be the most effective way of modifying their behaviour in favour of achieving climate stability.

• that governments pay up only when a stable climate has been achieved—any risk of failure or of undershooting the climate stability target is borne by Bondholders, rather than taxpayers.

Achieving a stable climate will almost certainly require a huge range of different projects. Reducing greenhouse gas emissions or sequestering carbon may be helpful, but they are not necessarily going to be the most cost-effective. Other ways including some yet to be discovered may be far cheaper. Kyoto is deficient in that it offers no incentives to discover and exploit these ways. Climate Stability Bonds would encourage the most efficient solutions given the knowledge available at any time, and they would stimulate research into finding ever more cost-effective solutions. This occurs because of the nature of the Bond mechanism, and requires no presupposition as to the optimal set of solutions. Governments, through the issuing body, would dictate only the objective—climate stability—not the ways of achieving it. Crucially too, the *objective* is one that all shades of public opinion can support. Without such support no coherent policy addressing climate change is likely to succeed. These and other advantages of Climate Stability Bonds are discussed in more detail in the next chapter.

Chapter 3

Climate Stability Bonds: comparison with Kyoto

This chapter looks in detail at the advantages of Climate Stability Bonds over Kyoto. Their main advantage is efficiency, and section A looks at some aspects of this. Sections B and C look at other complementary advantages of a Climate Stability Bond regime: the stability and transparency of policy goals. Sections D looks at the Bonds' more politically appealing money flows, and Sections E and F compare Climate Stability Bonds with two other 'more market' approaches: tradable pollution permits, and the contracting out of a desired outcome.

A Efficiency

The main advantage of Climate Stability Bonds over Kyoto is that, because they would inject self-interest into all stages necessary for achieving the desired outcome, climate stability, they would be *more cost-effective* than greenhouse gas emission cutbacks. Climate Stability Bonds give people the incentive to be efficient in achieving all processes necessary to reach the targeted outcome. Because stabilising the world's climate is certain to be a costly task, this cost-effectiveness is of critical importance. The remainder of this section looks more closely at how the Bonds' efficiency will influence three processes in particular:

1. investigating new activities,
2. responding to local effects and changing circumstances, and
3. costing the climate stability objective efficiently.

1. Investigating new activities

The Bonds specify and reward outcomes; they do not prejudge how these outcomes should be achieved. For a broad objective, such as climate stability, it may be preferable to encourage a full range of activities than to specify in advance, and with only current scientific knowledge, how the objective shall be achieved. Ideally, policies should not discourage research into, and application of, new, more efficient solutions than those that can currently be envisaged by policymakers. A policy such as Kyoto, that seeks to constrain certain prescribed activities and reward other prescribed activities, may not be optimally efficient, because our knowledge of the results of these activities, and of ones yet to be discovered or investigated, is changing all the time.

Climate Stability Bonds are not incompatible with the science underlying Kyoto. If bondholders' incentive-driven research leads them to believe that cutting back greenhouse gas emissions is the most efficient way of achieving climate stability, then that is what they will pursue, but on the basis of cost-effectiveness, rather than through being compelled to comply with a top-down approach that incorporates the very limited scientific knowledge that exists today. Cost-effectiveness in carrying out the existing envisaged way of attenuating climate change may be important, but so too is the fact that the Bonds will encourage people to investigate new activities that may be better at stabilising the climate, or mitigating the worst effects of climate change.

Many scientists, technologists, engineers and biologists, in countless research bodies the world over, are looking at the causes and effects of climate change and how to deal with or mitigate them. But there is no overall mechanism in place to allocate funding for these bodies on the basis of their likely cost-effectiveness. Within some countries, cost-effectiveness may be one criterion used to allocate funding. But in general, funding for a particular body depends on a range of factors, including ones that have little relevance to efficiency, such as: the body's existing size or existing level of funding; the number of people it employs; its contribution to the local economy; its fundraising skills; or the charisma of its public relations directors and their relationship to important politicians or celebrities.

Under the current regime, funding for climate change research projects has to compete with demands from a wide range of other government expenditure items. These projects may well be—indeed, most probably, are—run by people with the highest integrity and scientific knowledge, and it is highly likely that *within each body*, funds will be allocated impartially and with a view to obtaining the best result for each dollar outlay. The problem is that there is no overall incentive that will ensure that funds are allocated *between* these bodies, or to the creation of new ones, with a view to *overall* efficiency.

It is unlikely, perhaps, that *increasing the planet's albedo*—its reflectiveness—by, for example, launching mirrors or reflective particles into orbit around the Earth orbiting, could reflect a significant proportion of incoming radiation back out again, and so play a large role in achieving climate stability. But such an initiative may well have a contribution to make, as one of a wide range of diverse projects, all aimed at achieving a stable climate. Kyoto, despite all its vast expenditure, will do nothing to investigate this possibility. With its sole target of reduced greenhouse gas emission levels, Kyoto leaves innovative solutions out of the reckoning when disbursing funds.

Or take another example: the development of **genetically engineered cyanobacteria** that can soak up atmospheric carbon dioxide and convert it into a raw material for biodegradable plastic. In the last few years scientists in Japan announced that they can make the cyanobacterium (*Synechococcus* sp.) produce up to 10 per cent of its dry weight as polyhydroxybutric acid (PHB). When PHB is joined in a copolymer with hydroxyvalerate it produces a biodegradable plastic. And the only raw materials that the altered bacteria need to produce the PHB are water and carbon dioxide. The scientists wanted to use the engineered organism to extract carbon dioxide from exhaust gases in factories—simultaneously reducing emissions of a greenhouse gas while making a useful product.[12]

Again, cyanobacteria are unlikely to be the sole solution to the climate change problem. The point is that it is entirely possible that a range of solutions will be found that will collectively be far more efficient than Kyoto. It is also entirely possible that climate stability will be greatly assisted by natural phenomena. There is potential for new scientific solutions and for natural phenomena to invalidate the Kyoto approach almost overnight. The effect of Kyoto then would be to continue to impose costs out of all proportion to any benefit. But *there is no provision for reducing these costs*, or for backing out of Kyoto obligations in the event our knowledge or circumstances change to the extent that Kyoto becomes ridiculously extravagant. The Kyoto approach ensures that the entire planet will continue to bear the cost of greenhouse gas cutbacks, regardless of whether these cutbacks are necessary to, or efficient in, achieving climate stability. Indeed it is not difficult to imagine circumstances under which greenhouse gas cutbacks actually increase climate instability. But Kyoto does not allow for that possibility.

The resources involved in trying to achieve climate stability via Kyoto are colossal, so its inefficiencies are going to consume significant resources that could otherwise be used to solve real social and environmental problems. Kyoto is going to conduct its climate stabilising activities as if science stopped in the year 2000. Climate Stability Bonds, in contrast, would allocate funding in such a way as to encourage the exploitation of new initiatives. Any resource savings would be potentially of great value to those most in need of government funds, or to the environment generally.

2. Responding to local effects and changing circumstances

The second point, the incentive to experiment with different activities in different regions, is significant in that, while climate change is a global problem, its effects are local. Unlike Kyoto, Climate Stability Bonds can target these effects, rather than their assumed causes. The Bonds can be made redeemable subject to the achievement of such outcomes as reductions in the numbers of people killed or made homeless by climate-related events.

Climate Stability Bonds that incorporate such outcomes into their overall objective would encourage investigation of local circumstances. Here the superiority of Climate Stability Bonds could translate directly into alleviation of human suffering on a significant scale. If the Bonds were to target a reduction in the numbers of people killed worldwide by storms then funds for such projects as sea defences, for example, would be allocated under a Bond regime to those areas where the most benefit per dollar outlay could be obtained. One result is likely to be that bondholders would invest more heavily in sea defences in densely populated poor countries like Bangladesh, rather than in equally vulnerable, but thinly populated prosperous countries. There is nothing to stop these countries using their own funds for such purposes, but it would seem proper that global funds should be used to maximise the benefit to the world's population.

Similarly, a Bond regime is readily adaptable to changing circumstances. An unexpected increase in the frequency of storms, for example, or a rise in the sea level that exceeded all projections, would, under a Bond regime, lead to a channelling of climate stabilising funds into alleviating the worst effects of these unanticipated events. Under Kyoto the massive global resources deployed supposedly to stabilise the climate through greenhouse gas emission cutbacks cannot be so responsive.

Kyoto will provide no means of reallocating resources either to those parts of the world most in need, or in response to changing events. All Kyoto's efforts will be focused on cutting back gas emissions, at its own pace, and regardless of:

- the differing effects of climate change on different parts of the world, and

- of events arising that would make a reallocation of resources more beneficial in achieving either climate stability, or a mitigation of its negative effects on plant, animal or human life.

3. Efficient costing of objectives

Many environmental objectives are difficult to value, and achieving a higher degree of climate stability is no exception. Climate Stability

Bonds share with conventional policy instruments the need for some estimate of the value to society of a specified objective. We saw in chapter 2, section B, that, unlike Kyoto, a Climate Stability Bond regime caps the *maximum* funding to be spent on achieving climate stability. But Climate Stability Bonds have a further advantage over Kyoto in that competitive bidding combined with tradability guarantees that the *actual* cost of achieving the targeted outcome is also minimised. And if bondholders fail to perform, the cost to the taxpayer is zero. In maximising the efficiency with which the outcome is achieved, the market for the Bonds is elegantly efficient in conveying information about the cost of achieving objectives and, crucially for policymakers, how this cost varies with time and circumstances.

To illustrate this, let us assume that we have an agreed measure of climate stability, a 'Climate Stability Index' (CSI) embodying several desired outcomes, and that our current climate stability is found to be 50 on this scale. Assume further that there is an agreed target figure for this Index of 100.

Say that an initial tranche of Climate Stability Bonds is issued that will become redeemable when the CSI has risen by 10 points to 60. Further, let us assume that 1 thousand Climate Stability Bonds are issued redeemable for $1 million each once the CSI has reached 60. The maximum cost to the body issuing these Bonds is then $1 billion (one thousand multiplied by one million). But if the Bonds, when issued, fetch $500 000 each, then the market is saying that it thinks it can achieve this objective for just $500 million. It doesn't say when it thinks it can achieve that objective, but that can be inferred from market behaviour and the market value of the Bonds compared with other financial indicators. But what if the Bonds sell for virtually nothing, and the market value of the Bonds fails to move from that floor? That would mean that the issuers had miscalculated: in the market's view there is no realistic chance of the objective being achieved for an outlay of $1 billion in the foreseeable future. The issuing body can respond in different ways:

- It can wait for new technology to arrive (whose appearance will be hastened by the Bond issue), or for circumstances to change in other ways, such that the market sees the objective as becoming more easily achievable, and the value of the Bonds consequently rises. Or

- It can issue more Bonds, with the same specification, also redeemable for $1 million each. It can do this in stages, gauging the market reaction to each new tranche of Bonds, which will tell the issuing body the maximum cost of achieving the objective.

Either way, the issuing body can be reasonably sure that it is getting a good deal, expressed as climate stability per unit taxpayer outlay. But the Climate Stability Bond market does not merely reveal the total cost of achieving the objective. It also indicates the marginal cost of achieving further improvements. Say the one thousand Climate Stability Bonds initially issued do actually sell for $500 000 each. This tells the issuers, or participating governments, that the present value of the expected maximum cost of increasing climate stability from '50' to '60' is $500 million. The issuers may therefore judge that it is well worth being more ambitious, and aim for a further increase in the Climate Stability Index to 70. It could issue a thousand additional Climate Stability Bonds redeemable when this new level of stability is reached. Each of these new Bonds would (probably) have an initial market value of less than $500 000, reflecting the (probably) diminishing returns involved in further increasing climate stability. The point is that, by letting the market do the pricing of the Bonds, the issuer is getting an informed view of the marginal cost of its objectives. So if the Climate Stability Bonds targeting the new CSI level of 70 sell for $400 000 each, then the maximum cost of achieving that objective becomes $1.1 billion, being equal to: $500 million (paid out when the CSI rose from 50 to 60) plus $600 million (paid out when the CSI rose from 60 to 70). So the marginal cost of an increase in the Climate Stability Index of 10 points is revealed to have risen from $500 million to $600 million. Should the issuer then aim for a further increase in the Index to 80? Under a Climate Stability Bond regime *it would have robust information about the cost of doing so.*

In reality, the Bond market will continuously update this sort of information. Say that improvements in technology, of the sort that might be stimulated by an initial Climate Stability Bond issue, mean that it becomes much cheaper to remove carbon dioxide from the atmosphere. How would the market react to such a development? Once the new technology's effectiveness became obvious, the value of all Climate Stability Bonds, whenever issued, would rise. Instead of being priced at $500 000 and $400 000, the two Climate Stability Bond tranches of our example might fetch $800 000 and $700 000, respectively, per Bond. The total cost to the issuing body of redeeming these Bonds will not change—it will remain $1.1 billion (though redemption will occur earlier). But the market is providing new information as to the likely cost of future improvements in climate stability. The market now expects increases of 10 in the Climate Stability Index to cost $200 million (from 50 to 60), and $300 million (from 60 to 70). The new technology has reduced the costs from $500 million and $600 million (respectively). So the cost of any further increase in climate stability will also fall, and by following market price movements we can gauge approximately by how much.

These figures are made-up and simplified, but they do indicate the role that a Climate Stability Bond market could play in helping the issuing body—and the taxpayers—decide on their spending priorities. Climate Stability Bonds are efficient not only in supplying climate stabilising services, but in pricing societies' climate stabilising objective, and for much the same reason: the people who make up markets gather and reveal more information than a handful of government agents, and they have an incentive to use it efficiently.

B. Transparency

Climate Stability Bonds will help make policy objectives more transparent. Under a Climate Stability Bond regime the combination of targeted outcomes constituting the desired level of climate stability is explicitly identified. There is an important political dimension to this: focusing on a set of identifiable outcomes would encourage constructive participation in the political process. This means that measures taken to achieve these outcomes will be more likely to attract public participation and support than activity- or institution- based measures. At least as important, the desired level of climate stability would be explicitly costed. This means that the maximum value that society wishes to place on climate stability will have to be decided and publicly known before any programmes have begun. Once that has been determined, the issuing body will be able to decide on the Bonds' redemption value and the number of Bonds to be issued. Costing outcomes in this way would make the tradeoffs between climate stability and other global goals more transparent. So it would also make people's expectations of what governments can achieve in both the environmental and social arenas more realistic.

It is important also that a clear expression of the desired outcome, as embodied in the Bonds' redemption conditions, would mean that progress toward them could be accurately monitored. Market prices and their movements would constantly reveal, to everybody, huge amounts of information useful to policy makers—including information about the expected costs of achieving the objective.

C. Stable goals

Achieving and sustaining a significantly higher degree of climate stability is likely to take many years. How is Kyoto likely to fare when more immediate and obvious emergencies, demanding scarce government resources, arise? Or when new science shows Kyoto to be either unnecessary, or grotesquely inefficient, or even counterproductive? The goal of climate stability is more stable than any presumed link between cause and effect. The majority of decision makers currently see cutting greenhouse gases as the best way of tackling climate change. But there is no necessary reason why this will always be the case. In this context it is worth repeating that there a possibility—and a possibility strong enough to discourage whole-hearted participation and compliance with Kyoto—that:

- science will lessen the relative significance of any link between greenhouse gases and climate change, or

- that climate stability will occur without any intervention, or

- that new events will deter governments from ensuring their countries' compliance with their Kyoto obligations, or

- that new governments will come to power who either assign Kyoto a lesser priority, or repudiate it.

Under a Climate Stability Bond regime, the stability of the *desired outcome* makes it unlikely that investors will be deterred from taking measures to achieve it just because of changes of this sort. This contrasts with Kyoto, where we have already seen governments, most notably the US Government, refuse to be bound by its targets. Significantly, these governments object not because they disagree with Kyoto's aims, but on the grounds that the proposed mechanisms are inefficient and costly (and unfair in that developing countries are not also obliged to cut their greenhouse gas emissions). Kyoto is vulnerable to changes in either scientific knowledge or political attitudes that would make other governments change their minds, and think along the lines of the current US administration. Under a Climate Stability Bond regime, it would remain the responsibility of bondholders to decide on the most efficient means of achieving the goal. Once the goal had been agreed to, the effects of any such governmental vacillation would be felt by the bondholders, via changes in the market price of their Bonds. The goal of climate stability would remain unchanged, and it would still be in bondholders' interests to achieve it.

D. More attractive money flows

A further advantage of Climate Stability Bonds is that, in many cases, they will have politically appealing money flows. Kyoto will inflict identifiable losses on certain people and countries in pursuit of its greenhouse gas-controlling objective. The scale of these losses has already led to the US weakening, perhaps fatally, what should be a global effort. Climate Stability Bonds, however, would reward people for achieving successful outcomes. The Bonds would of course be redeemed by funds from governments' general taxation revenue and taxes would still have to be levied to provide this, but there is, nevertheless, a presentational advantage.

The other, highly significant, money flow advantage of Climate Stability Bonds is that taxpayers will incur expenditure only when the targeted improvements in climate stability have actually been achieved. For this reason, the Bonds may attract greater political support than Kyoto, with its enormous, financial (and political) costs, payable in advance.

E. Comparison with tradable permits to pollute

A tradable permit regime specifies the maximum amount of pollutant that can be discharged. It then issues tradable permits to emit amounts of pollutant making up this total. Markets decide the price and allocation of these permits. In the United States, for example, markets for permits to emit acid-rain producing sulphur dioxide have been in operation for several years. Tradable permits are especially useful in allocating unpriced resources, such as the assimilative capacity of the environment, and also for targeting pollutants that have marked thresholds.

Kyoto will embody rules and guidelines for "market-based mechanisms" that allow flexibility to participating countries in meeting their obligations. These include emissions trading, joint implementation, and the Clean Development Mechanism. Amongst the critical operational issues to be considered are 'transparency' in making it possible effectively to track emission units, and key measurement, reporting and verification issues.

Emissions trading means that if countries' own pollution levels remain too high, they can trade for credits from countries that exceed their targets. The trading rules have yet to be established, but is already clear that carbon dioxide will be the key commodity. One tonne of carbon dioxide emissions (or the equivalent of another of Kyoto's greenhouse gases) is commonly measured as one tradable emissions allowance. These allowances, which may at some point become government-sanctioned credits, can change hands in a variety of ways. In the developed world, if country A can meet its emissions-reduction target without using up all of its credits, it can sell the remaining pollution rights to country B. In addition, developed nations can earn credits by helping other developed nations lower their emissions. Another provision would allow developed countries to earn credits by working to cut pollution in developing countries. One key idea is to encourage green habits and best industrial practices to spread across borders. A United Nations convention on climate change is scheduled to sort out the practical details of the trade at a Marrakech summit late in 2001.[13]

How do the market mechanisms contemplated by Kyoto measure up against a Climate Stability Bond regime? First, and at the risk of being repetitive, is the important fact that Kyoto's policies have as their objective the cutting back net emissions of greenhouse gases, rather than achievement of climate stability. So even a highly efficient system emissions trading system will be efficient only in that limited sense.

Second is that emissions trading schemes require large amounts of information if they are to effectively tackle broad objectives. The United States' tradable permit programme designed to reduce sulphur dioxide emissions is successful because it is easy to monitor and enforce. Most of the sulphur dioxide emissions came from just 2000 smokestacks in the American Midwest. But emissions of most greenhouse gases result from many sources and many different processes: carbon dioxide, for example, is emitted by cars and homes, while about 8 per cent of greenhouse gases in the developed countries are emitted by the agriculture sector.[14] Immense quantities of information would be needed to establish, monitor and enforce a comprehensive system of carbon dioxide emission controls using tradable permits to pollute. Similarly with some of the other greenhouse gases. A Climate Stability Bond regime, however (assuming bondholders thought reducing greenhouse gas emissions were an efficient way of tackling climate change), would simply target greenhouse gas pollution levels, and let bondholders decide how best to achieve them.

In reducing emissions of some greenhouse gases there may be a role for tradable permits, which can work well with intrinsically large-scale processes. Such processes can be monitored quite easily, because there would be no fear that doing so would lead to offsetting increases in pollution via the setting up of difficult-to-monitor small-scale processes. But most greenhouse gases do not fall into this category.

Third: the role of incentives and self-interest under Kyoto is likely to be less pervasive than under a Climate Stability Bond regime. Under Kyoto, markets will be set up for the tradable emission allowances of all participating countries. Each country will be able, through its dealings with other countries, to organise its emissions and removals activities efficiently, *as a country*. But *within* a country, it is only possible, not certain, that governments will establish a system that injects market incentives into all necessary stages leading to the achievement of the country's greenhouse gas objective. Tradable permits to pollute may be one climate policy instrument, but there is a large portfolio of alternatives from which governments can choose. They include: taxes on emissions, or on carbon or energy use; (non-tradable) permits; the provision, or removal of subsidies; the imposition of technology or performance standards or energy mix requirements; product bans; voluntary agreements; government spending and investment; and support for research and development. In choosing a mix of these policies, and in the implementation of most of them, it is unlikely that efficiency will play a prime role. Each government will have to comply with its international obligations to cut (net) greenhouse gases emissions, but it will have little incentive to do so with maximum efficiency: as in other policy areas, political expediency is likely to influence government behaviour.

A Climate Stability Bond regime, on the other hand, would ensure that efficiency, measured as the increase in climate stability achieved per dollar outlay, operates at every level of aggregation, including within every participating country. As well, bondholders would probably allocate their climate stabilising priorities *between* countries differently than does Kyoto. Again, efficiency would be their guiding principle.

Fourth: while the details of Kyoto's emissions trading system have yet to be devised, it is likely that they too will embody the scientific relationships assumed or known at the time of formulation. Kyoto in general suffers from these uncertainties over the relative contributions that greenhouse gases make to climate change. A market-based system allowing sinks to offset greenhouse gas emissions would suffer from the even greater uncertainties in the relationships between these sinks and atmospheric greenhouse gas levels. It is likely that any trading scheme would fix these relationships according to today's science: so, for example, a given area of a particular type of forest would be regarded as equivalent to emissions of a fixed level of carbon dioxide equivalent, regardless of any new knowledge that will come to light about the effects of these elements on climate change.

A Climate Stability Bond regime would begin with the same uncertainties, but bondholders would have a powerful and *continuing* incentive to investigate the relationships between all these activities, and to *adapt to new knowledge*, always with a view to achieving the goal of a more stable climate as efficiently as possible.

F. Contracting out

Climate Stability Bonds have the effect of contracting out the *achievement* of climate stability to the private sector, while the backers of the Bonds, most likely the world's governments, set the objective, and undertake to redeem the Bonds. It has similarities with the more conventional contracting out, which usually involves government specifying the outputs it requires, in terms of the nature and level of service required, and inviting the private sector to bid for the contract to supply these outputs. Specifying outputs may be helpful when there is a definite relationship between a small number of outputs and the desired outcome. But for many objectives, including climate stability (and other environmental goals), such relationships are many and obscure. Contracting out of output-supplying services generally means that the required output must be specified to a high degree. This imposes its own costs, and means that contracting out tends:

- to be limited to particular stages of outcome-delivery; and

- to reinforce established ways of doing things.

Of course, government could contract out *outcome*-supplying services. (In the late 1980s the New Zealand Government did in fact contemplate this.) Doing so would provide some of the advantages of Climate Stability Bonds, but not all. What distinguishes a Climate

Stability Bond regime from the contracting out of outcomes is the Bonds' tradability, which has three main effects.

1. The possibility of trading the Bonds avoids the problem of possible collusion (tacit or not) between bidders for contracts. Under conventional contracting out mechanisms, inflated bids can succeed if the bidders agree (explicitly or not) to inflate their bids. Under a Climate Stability Bond regime, the Bonds could be bought and held by anybody, not just the people or companies already involved in carrying out the target-achieving projects, or well set up to do so. So the possible bidders under a Bond regime would not be limited to a few likely operators but would encompass all who are prepared to do, or to finance the doing of, projects that will help achieve the objective. The fact that anybody can be involved in the bidding for Bonds will discourage people from making excessive bids, so ensuring that the targeted objective can be achieved as cost-effectively as possible.

2. Perhaps more important that the Bonds' tradability encourages suppliers of services to *continue to minimise costs* and maintain efficiency, even after they have begun to achieve the targeted goal. Under current contracting out arrangements there may be a tendency for contractors, or their employees, having won a contract, not to maximise the speed and efficiency with which they go about solving the targeted problem. While they can sometimes benefit from being efficient, they cannot always enjoy this benefit in terms of immediate cash capital gains. And though there is scope for incentive payments, or penalty clauses, these are crude, ad hoc arrangements, which are costly to administer and impose. Under a Climate Stability Bond regime, if bondholders are unexpectedly efficient (or if external events

are unexpectedly helpful) they can sell their Bonds and realise their capital gains before all the necessary work has been carried out. And if bondholders turn out to be inefficient and insist on holding onto their Bonds, they would be the losers, not taxpayers.

3.   Climate Stability Bonds' tradability transfers the risk of breach of contract from the taxpayer to bondholders. If, under conventional contracting out systems, the successful bidders fail to do what they were legally obliged to do, then it is up to the aggrieved party—the central or local government agency—to take proceedings against them. Even if such actions win some form of redress, it is rarely sufficient compensation, and proceedings can be protracted and costly. Under a Climate Stability Bond regime, underperforming bondholders will find a ready market for their Bonds in people who believe they can be more efficient.

Chapter 4

Practical aspects and potential pitfalls

At first sight, a Climate Stability Bond regime may seem outlandish: it means that governments hand over responsibility for accomplishing an important global environmental objective to the private sector. It is true that a Bond regime would represent a radical change in the way in which society sets about achieving its goals. But it is important to realise that under a Climate Stability Bond regime governments are merely contracting out the *achievement* of a stable climate. Governments, acting in concert, in consultation with scientists, and through the issuing body, will still set the climate stability goal and, by undertaking to redeem the Bonds, will still be the ultimate source of finance for the projects that achieve it. Nevertheless, there are some practical aspects, and potential pitfalls to consider. This chapter begins, in section A, by looking at the incentive Climate Stability Bonds may give to commit negative acts, including 'insider trading'. Sections B and C, look, respectively at effects on government's behaviour, and at how government will interact with a Bond regime. Section D, looking at how a transition to Climate Stability Bonds can be effected, ends the chapter.

A. *Incentive to commit negative acts*

Climate Stability Bonds work by giving financial incentives for people to achieve a particular environmental goal. Realistically, however, they may also encourage people to break the law to do so, or to commit acts that while not illegal, conflict with society's other environmental or social objectives. Acts of this kind may already be illegal in some jurisdictions, but before issuing Climate Stability Bonds governments should be aware that there would be greater inducements to commit them.

Examples of such negative acts are:

- emitting pollutants that have a stabilising effect on the Earth's climate but that have a negative effect on human, plant, or animal life, or on the environment generally. Bondholders may emit these pollutants in countries where it is not illegal to do so, but they would also have the incentive to do so where such emissions are controlled or illegal.

- falsifying data on one or more components targeted by an issue of Climate Stability Bonds.

Much would depend on how the target of a stable climate is specified. For example, one targeted variable could be the temperature, as measured at a selection from a large number of sites around the world. Unscrupulous bondholders might try to influence the selection of sites to suit their purpose. In such a case, vigilance, integrity and judicious specification of the targeted objective would help.

It is likely though, that even a carefully specified climate stability objective will not always eliminate or mitigate the kind of negative-but-legal, or illegal, activities that the Bonds may stimulate. So how can this potential problem be solved? The solutions have to do with the way in which the Bonds are introduced, and with the role of government.

A cautious, gradual, introduction of Climate Stability Bonds is one means by which adjustment problems can be minimised. Initial issues can target small increases in climate stability, with the issuers being careful to watch for any negative behaviour and anticipate it in subsequent Bond issues. As well, the number of Bonds issued, and their redemption value, could be designed so that their initial market value would be very high, limiting ownership to major concerns, whose behaviour could be more easily monitored. If, despite such precautions, bondholders behaved illegally, government could prosecute the perpetrators. If bondholders behaved in negative, but legal ways, government has other options. In ascending order of severity, government could:

- persuade or cajole bondholders into toeing the line. It could do this publicly or privately—initially, at least, bondholdings would be registered in the same way as shares;

- buy back Bonds, which would have the effect of lowering the market price of Bonds remaining on the market;

- legislate against the unforeseen activity; or in extreme cases

- declare the Bonds null and void, and offer compensation related to the Bonds' issue price, or their current market price.

The issuing body, in conjunction with the world's governments, will also have to organise reliable and accurate monitoring of climate change so that progress towards the attainment of climate stability can be impartially assessed. This monitoring must be seen to be independent of bondholders, who could benefit unfairly from dubious data collection, and who will have their own incentive to gather similar information. Naturally the information as to how close the objective is to being achieved will have value. It is not difficult, for instance, to imagine international temperature statistics being sought in advance of official publication and used for 'insider trading' purposes. If too much insider trading went on, it would increase the riskiness of Climate Stability Bonds to those without access to this information and so tarnish their value as an investment. So how can insider trading be minimised?

• In many cases data gathering and collation would have to be more transparent. There would be a ready market for this information, which would help to make these processes more robust. There would be more interest in more frequently updated information, so that progress toward achieving objectives could be more readily charted. If large sums of money were at stake, there would be a great deal of private information gathering: investors, bondholders, and financial commentators will be taking their own soundings during the lifetime of each Bond issue. All this would serve to keep the official information assessors honest.

• Those involved in gathering, collating and processing information would be bound by terms deterring or forbidding them from abusing privileged information.

• Indicators for targeted objectives could be chosen with a view to minimising the possibility of insider trading being an important factor. Combinations of indicators could shorten the length of the information chain. The same effect could be achieved by the issuers' stipulating that the Bonds would be redeemed on the basis of data from a random sample of sites, rather than from all sites or any pre-arranged subset of sites.

- The objective itself could be chosen to minimise the possibility of insider trading. Bonds targeting a long-range objective such as halving climate instability rather than cutting it by 10 per cent, would probably be less sensitive to insider trading. With long-range objectives, each datum illegally withheld from the Bond market would probably represent a smaller proportion of the total relevant information available to that market, and so have a lesser effect on the Bond's market value.

None of these ways of mitigating insider trading will always be fully effective. That said, there are already sensitive indicators, such as unemployment or retail sales figures, that are capable of moving markets, and so there are already in place mechanisms to control access of such information until it is time for publication. There are also sanctions against those who obtain, and act on, such information illegally. These mechanisms and sanctions might need to be strengthened under a Climate Stability Bond regime, but it remains to be seen how important abuse of insider information will be. While insider trading would mean that unscrupulous people benefit at the expense of the public, it does not generally impede the operation of markets. Markets continue to function and the possibility that a low level of insider trading goes on is generally discounted by the broader market.

Unanticipated negative acts in general, and insider trading in particular, are most likely to occur when Climate Stability Bonds are first issued. Lessons would be learned to be incorporated into later issues. These lessons would extend beyond how to deal with bondholders' behaviour. They might, for instance, give some direction as to how best to specify elements of the overall climate stability objective, or how the Bonds can most usefully supplement existing policy. National governments could collate their experiences, and share any lessons learned with the issuing body.

B. *Effects on government's behaviour*

Another possible problem arising from the integration of Climate Stability Bonds into the current policy-making system arises from government's role as creator of statutes. Government can pass laws affecting the Bond price, or its actions could influence the Bond price in other ways. For instance: particular governments could come under great pressure to increase the taxes it imposes on fossil fuels from holders of Bonds who believe that such lobbying would be the easiest way of ensuring a rise in the value of their holdings. Note, though, that the source of the pressure, and the motivation for it, would be easy to identify. And note also that lobbying is a legitimate activity, and that its results need not be negative. There is no reason why bondholders, in common with other pressure groups, should not lobby politicians. They would be doing so mainly out of financial self-interest of course. But other pressure groups are also self-interested, and in the case of bondholders their self-interest is, or should be, congruent with society's interests. Bondholders would lobby for legislative change, and they will benefit in obvious, financial ways. Potential investors in the Bonds will take into account the likely influence of bondholders on legislation, and the potential influence of changes in legislation on the speed at which climate stability is achieved, when they assess the value of the Bonds.

These influences make it important for there to be some element of consultation when defining targeted objectives. People do become rich by using their influence on politicians under the current system, but they are less identifiable, and they do so in ways that are not always easy to identify. Politicians will have to weigh the evidence for and against any course of action promoted by lobbyists, with due regard to the lobbyists' motivation. In the end, though, it is up to potential investors in Climate Stability Bonds to take into account likely or possible changes in the legislative environment when bidding for the Bonds.

The threat of bondholders lobbying governments for legislative changes can have a positive aspect. For Climate Stability Bonds to be as successful as possible, governments would ideally give assurances as to their future behaviour. So, for example, prospective purchasers of Bonds targeting water pollution may want to know governments' plans to target fossil fuels, or to subsidise renewable energy, or even to impose speed restrictions on motorways. Governments would maximise interest in the Bonds by being as open as is practical about its legislative and spending intentions as soon as possible. They could also undertake not to do such things as eliminate taxes on fossil fuels. When deciding how much Climate Stability Bonds are worth potential investors would, of course, regard all such assurances with an appropriate degree of scepticism.

Of course, if an initial tranche of Climate Stability Bonds targets only a small increase in climate stability, then governments' long-range plans will not be so significant to prospective bondholders. Targeting incremental improvements in climate stability, it may emerge after trials of the Bond concept, could be the best way of dealing with any uncertainties over future government behaviour. Or it may be that, in most cases, these uncertainties turn out to be a relatively insignificant component of the total uncertainty that attaches to investment in Climate Stability Bonds. Markets routinely deal with low levels of risk and uncertainty by assigning lower values to riskier financial instruments.

Is there a case for preventing national governments from trading in Climate Stability Bonds? There are at least two aspects to this question.

1. As creators of statutes governments are bound to possess privileged information. If they held a significant number of Bonds they might also be tempted to time their lawmaking or spending activities to maximise their Bonds' capital appreciation. On the other hand, any changes in behaviour could be anticipated, and discounted, by potential bondholders.

2. National governments could also participate in the Climate Stability Bond market as competitive suppliers of objective-achieving services. They could buy Climate Stability Bonds in their capacity as active investors. Unlike in industry the private sector would be unlikely to cry 'unfair competition', even if the operations of these agencies were heavily subsidised, because its own Bonds would appreciate as a result of the government, or government-inspired, activity.

The solution is probably twofold. First, governments' trading activities ought to be fully transparent. Second, if government agencies are to participate in the Bond market as suppliers of climate-stabilising services they should be denied privileged access to information. It would be preferable that any profits these agencies make on Climate Stability Bonds, or losses that they incur, should accrue to that agency. The people who work for the agency must have the same incentives as the private sector bodies to perform efficiently. This would change the character of these agencies, and would probably lead to their ultimate divorce from the public sector.

C. Existing institutions and the transition to a Climate Stability Bond regime

Hundreds, probably thousands of scientists, research institutes, academic and government bodies, are already researching the causes and consequences of climate change, and ways of mitigating climate change. How would their funding be altered by a Climate Stability Bond regime? Much would depend on whether resources allocated to redeeming Climate Stability Bonds were additional to current funding.

Few bodies charged with achieving climate stability are currently financed in ways that correlate directly with their performance. Nevertheless these bodies are the main repositories of expertise for solving environmental problems, and some of them are bound to be efficient, or to be capable of becoming efficient. It would be unwise as well as unfair and unnecessary to divert their funding at short notice into funds allocated to redeem Climate Stability Bonds. The answer, at least for those institutions whose funds derive largely from government, is a gradual transition. Such a transition to an outcome-based, rather than institution- or activity- based funding programme, would see these bodies' state funding gradually decline, while expenditure allocated by bondholders on the basis of their expected contribution to the targeted outcome—a more stable climate—would gradually rise.

One mechanism would be for governments to reduce their funding of climate-related programmes and institutes by 3 per cent a year, in real terms. (They could use these savings to contribute to the Climate Stability Bonds' redemption funds.) After five years, each programme or body would be receiving directly from its government only 85 per cent of what it formerly received. But bondholders may choose to supplement the income of some of these bodies. They may judge some of them to be especially effective at converting the funds they receive into measurable climate stabilising benefits, as defined by Climate Stability Bonds' redemption terms. It may be that the most effective bodies, in terms of achievement per unit outlay, are working in especially vulnerable areas, where small outlays would probably bring about larger benefits in terms of, say, the number of lives saved per weather-related adverse event. Or bondholders may judge that a particular research body is worthy of additional funding, because they assess its research as being most likely to yield climate stabilising measures that will be efficient in converting its funding into effective climate stabilising programmes. For such bodies, bondholders could supplement government funding. It may well be that these favoured bodies end up receiving annual sums amounting to considerably more than 100 per cent of their former income.

It is possible that investors in Climate Stability Bonds will look at completely new ways of achieving their objective; ways that currently receive no, or very little, funding. To give a not entirely unbelievable example, they may be convinced that one of the best ways of achieving society's long-term climate stability goal is to increase women's access to family planning information and resources, in accordance with the agreement reached at the end of the Cairo Population Summit of 1994.[15] Following this logic, they may find that one of the most efficient ways of doing this is to support non-governmental organisations, including registered charities, that already have this objective, or to lobby government and religious authorities in some countries, again to encourage compliance with the Cairo Agreement. It is difficult to imagine how any policy other than Climate Stability Bonds, still less Kyoto, could even investigate, let alone implement, such an indirect approach, however effective it may be shown to be.

Could Bonds targeting remote objectives, such as a much higher degree of climate stability than currently pertains, be compatible with a gradual transition of the type described above, during which funding to existing government-funded institutions falls by 3 per cent annually? At first sight there is an apparent mismatch between small reductions in government spending and the large sums that would be needed to motivate potential investors in Bonds that target remote objectives. The critical point here is that bondholders will be investing not on the basis of the annual reductions in government expenditure on existing institutions, but on the basis of the redemption value of all the Bonds issued. To be more precise, it will be this total redemption value, minus the Bonds' existing market value, that will drive bondholders' investment decisions. This sum could be many times each year's incremental reduction in a particular body's state funding. One of the virtues of a Climate Stability Bond regime is that even in the short term bondholders would begin to invest in projects with a long range objective, on the expectation of capital gains that might arise only in the distant future.

It may be possible to expand spending allocated via the Bonds at a faster rate than the 3 per cent suggested: expertise in the environment is still relatively mobile, making it easier quickly to establish new outcome-based institutions or to reorientate existing ones.

Note that, while changes in the source of funds would be gradual, those involved in existing institutions with climate stabilising intentions are likely to react by quickly reviewing how *all* their existing programmes and projects operate. If bondholders see existing institutions and programmes as being particularly effective in achieving climate stability, then they will be inclined to invest in them. On the one hand, the switch in funding tells existing institutions that they can expect to see their relatively ineffective operations receive diminishing funds in the future. On the other hand, their effective operations can look forward to higher—possibly much higher—funding. A gradual transition could entail 3 per cent annual cuts in state funds to existing institutions.

But this would be balanced by the likelihood of additional income if institutions could impress Bondholders with the cost-effectiveness of their climate-stabilising projects. They may have to devote some of their resources into persuading bondholders of the cost-effectiveness of their activities; but this would not represent a radical difference from the way many of these bodies currently lobby for government funding nowadays. Under a Bond regime, though, they would do their lobbying on a more transparent, outcome-oriented, basis.

D. Opposition from existing institutions

The main opponents of a Climate Stability Bond regime are likely to be those people who have put a lot of time and energy into negotiating Kyoto. A huge industry is likely to develop around Kyoto. It will not be limited to the bureaucracy of regulation, surveillance and enforcement, but will also see the channelling of scientific expertise into the technology of reducing greenhouse gas emissions and removing greenhouse gases from the atmosphere. Sadly too, Kyoto has acquired symbolic significance amongst many people, non-governmental organisations and governments for whom its meaning transcends its capacity to achieve its stated purpose, and opposition can be expected from them.

It is also possible that there will be opposition from politicians and civil servants in many countries, as they contemplate their discretion to make decisions on climate change programmes passing out of their control and into the control of bondholders. They might also see that the underlying principle of Climate Stability Bonds could be applied in areas other than climate change, which would threaten the public sector's control over spending on such significant concerns as health, education and welfare. They may also begrudge the power of the body that issues the Bonds (see box).

The issuing body

The power of the body issuing Climate Stability Bonds may cause resentment in some governments or non-governmental organisations. The issuing body may be a new agency as an agent of, or under the auspices of, the United Nations, the World Bank, or the Intergovernmental Panel on Climate Change. Its powers would be:

• To specify Climate Stability Bonds' targeted objectives. This process, while it would no doubt entail complicated and difficult negotiation would, most likely, be less divisive than Kyoto. Simply: there will be far more agreement about the sort of climate the planet needs, than about the best way of achieving it. But even when the ultimate desired degree of climate stability is known, the issuers will have to decide on the optimal approach from a wide range of possible approaches. For example, would one large single issue of Climate Stability Bonds be preferable to a series of Bond issues targeting intermediate stages?

• To oversee the issue and redemption of Climate Stability Bonds. The issuers may, at least until the Bond principle has been tried and tested, also want to set up a system of registration and tracking of the Bonds, so as to help trace any negative activities that the Bonds might stimulate.

• To collate and present data from independent scientific bodies on the climate, specifically data that measures progress toward the achievement of targeted climate change objectives. The Bonds might encourage the falsification or manipulation of climate-related data, and it would be up to the issuing body to ensure that its data were as free as possible of controversy.

Public sector trade unions can be expected to resist Climate Stability Bonds, as they generally have opposed privatisation or voucher schemes, or indeed anything that they believe (or say they believe) opposes the narrow interests of their members and the status quo. Their reasons would be similar to those of some politicians who may also oppose Climate Stability Bonds despite the likelihood that the Bonds could achieve their stated objective more efficiently. This opposition would come from the unions' natural desire to hold on to power, or for purely ideological reasons arising from their apparent belief that public control of the means of supplying services is an end in itself.

However, support for Climate Stability Bonds could come from open-minded economically literate non-governmental organisations, and from all those who are sincere in their wish to see improvements in the position of the poorest members of society—many politicians do fall into this category. The efficiency benefits of deploying Climate Stability Bonds rather than Kyoto would mean huge savings for taxpayers.

So the poor, and those who claim to represent them ought to support the Bonds. The Bonds though will create incentives for people to regroup with the common aim of achieving climate stability. These groups will have their own dynamic and while their overriding interest will not necessarily *be* the public interest it will, provided scrupulous attention is paid to the Bonds' redemption terms, certainly *coincide with* it. The effects of creating such interest groups will create a powerful momentum in favour of the Bond principle.

Chapter 5

Possibilities and limitations

In mid-2000 research published in *Nature* sought to dispel fears that forests may play a catastrophic role in the process of global warming. Two separate studies indicated that the link between carbon dioxide output and temperature is both more complex and less alarming than had been previously thought. In one of the studies United States scientists found that long term decomposition of organic matter was unaffected by short-term swings in temperature. 'It means that forests are service as a carbon sink...providing a global environmental service by removing carbon dioxide from the atmosphere.'[16] The researchers said that it was time to recalibrate computer models, so that the new data can be taken into account. Indeed.

Studies such as these are published almost routinely. They are continually obliging us to revise our assumptions about the extremely complex science of climate change. But Kyoto's overarching objective, it is worth emphasising, is not to achieve climate stability. It is to reduce the net emissions of what are today considered greenhouse gases.

A Climate Stability Bond regime would be radically different from Kyoto. The Bonds would work by injecting market forces into the achievement of climate stability. Targeting the desired outcome will provides a continuous incentive for investors in the Bonds to investigate diverse solutions and resolve scientific uncertainties as cost-effectively as possible. Kyoto focuses exclusively on controlling those gases that (quite probably) contribute to climate change, but it does so on the basis of 1990s science. It fails to use market incentives, so even in its own narrow terms it will most probably fail to allocate resources rationally between greenhouse gas emission controls and measures that remove these gases from the atmosphere. Worryingly too, Kyoto is difficult to adapt to our rapidly expanding scientific knowledge of the causes and effects of climate change.

Despite all its theoretical advantages, however, and the fact that it has been in the public arena for some years (see bibliography), the Bond concept has not yet been applied in any sector, and it may have to overcome opposition before it progresses further. It is true that under a Climate Stability Bond regime, governments acting collectively would continue to set and rank their climate change o. But they would surrender, within limits, their power to dictate *how* objectives shall be achieved, and which institutions shall be charged with achieving them. Governments, along with the public servants who administer Kyoto, will be reluctant to lose this power. Also, beneficiaries of some current policies, particularly those with obscure goals, may oppose a shift toward more transparent policymaking.

For these reasons, and because it is a new policy instrument, the Bonds will need to be introduced cautiously. National governments could first try out the Bond concept on national environmental objectives. Or globally backed Climate Stability Bonds could be used alongside Kyoto. Rather than aiming for climate stability, early Bond issues could target 'the number of people worldwide killed or made homeless by adverse weather'. Such a contained, easily identifiable, goal would help the Bonds gain acceptability amongst the public, and encourage policymakers to adopt the concept and apply it more widely.

And Climate Stability Bonds need not be deployed in isolation. They can, if deemed too radical to be the sole climate stabilising policy instrument they can be used to complement Kyoto. If even that is thought to be risky, the Bond principle could be adapted to the narrow Kyoto objective: instead of Climate Stability Bonds, governments could issue 'Atmospheric Composition Bonds', which would become redeemable when the gases making up the global atmosphere conformed with some desired composition. Of course, such an objective might not result in a more stable climate. Like Kyoto they may be completely ineffective in changing the climate in any discernible way. But because they would inject market incentives into controlling net greenhouse gas emissions they at least they would be cheaper than Kyoto.

Efficiency

Climate Stability Bonds would harness self-interest into the achievement of a more stable climate, and ensure that this goal is achieved as cost-effectively as possible. What even the Bonds cannot do is answer to question of whether striving to achieve a more stable climate is a sound use of scarce resources. Opponents point out that Kyoto will cost an estimated $150 billion a year, and possibly a lot more, and that, according to Unicef estimates, just $70-80 billion a year would ensure that developing countries have access to health, education, water and sanitation.[17] The case for targeting only the negative effects of climate change would be stronger—except that not all these negative effects are currently identifiable. For some proponents of Kyoto though, efficiency is not the major issue. The IPCC, in one of its 2001 reports, proposes a radical change in lifestyles, exhorting us to share resources, choose free time instead of wealth, quality instead of quantity and 'increase freedom while containing consumption.'[18] This is an unfortunate indication of how a readily a political agenda can displace the original question—how to deal with climate change efficiently—even in a body whose stated purpose is to gather scientific information about climate change. It shows the degree to which Kyoto has totemic significance, and points to the intensity of debate and opposition that a concept such as Climate Stability Bonds is likely to face. Kyoto is in danger of becoming similar to 'recycling' (see box): something that is believed in by the devout, but which does little to achieve its supposed objective.

Recycling: a means to an end or an end in itself?

It is not only with respect to climate change that the principle of targeting the desired outcome, as embodied in Climate Stability Bonds, has advantages over the conventional activity-based approach. Take one environmental 'objective' that is often promoted: 'increasing the proportion of household waste that is recycled'.[19]

There is real confusion here. 'Recycling' has been taken to be an end in itself, rather than a means to an end. If the actual aim is to improve the environment, then it is not at all certain whether more recycling will do that. There are environmental costs in recycling too, and for most products and in most circumstances it is an open question whether or not recycling something has a bigger negative impact on the environment than using the same thing just once. Recycling a glass bottle, say, means using additional energy to clean and sterilise the bottle. It also means manufacturing, transporting and disposing of the necessary detergent. To show that recycling is better for the environment one would have to show that all the envi-

---

ronmental costs of these, and additional processes, are lower than the environmental costs of disposing of the old bottle and manufacturing a new one. This would be a difficult thing to prove, even at a single point in time. And technologies, and therefore environmental costs and benefits, are constantly changing. They also differ according to such variables as location. Some households, for example, may be sited right next to a recycling plant and that could be enough to tip the environmental balance in favour of recycling—for those households. Few 'life cycle analyses', which compare all the environmental costs and benefits of recycling with the alternative, have been done, and those that have do not show unequivocal evidence in favour of recycling on environmental grounds. One life cycle analysis estimated that the manufacture of paper cups consumed 36 times as much electricity and more than 500 times as much wastewater as the manufacture of much-derided polystyrene foam cups.[20] Another study found that while disposable nappies create around two times as much trash by volume as recyclable cloth nappies, they are probably more friendly to the environment, because their production consumes less energy and 50 per cent less water than that of cloth nappies. They also generate 40 per cent less air pollution, and 86 per cent less water pollution.[21]

- Climate Stability Bonds, instead of targeting the alleged means to an end, target the desired outcome itself. In the same way, rather than have as a policy goal recycling, whose net effect on the environment is unknown, it would be far better to target environmental pollution directly. A Bond regime could do this by targeting some or all of a combination of pollution indicators, which might include:

- nationally averaged indices of pollution, weighted according to lethality, taken at sites throughout the country, and comprising proportions of noxious air and water pollutants;

- numbers of people found in surveys to be suffering from pollution-related ailments; or

- responses of residents and overseas visitors to questions about their perception of environmental pollution.

Possibilities: applying the Bond principle more widely

Could the Bond principle be deployed in other policy areas? In fact Climate Stability Bonds are only one possible application of the principle of Environmental Policy Bonds, devised to inject market incentives into the solution of a range of national and global environmental problems. These, in turn, complement Social Policy Bonds, intended to inject market incentives into the solution of social problems, such as crime, unemployment, illiteracy and poor health standards. The bibliography cites references to papers describing these further applications of the Bond principle, and an edited version of a paper outlining the concept and its wider application appears as the Annex to this book. So far no government has issued Social or Environmental Policy Bonds but in many ways it would be easier, though perhaps less urgent, for governments to apply the Bond mechanism to such domestic social and environmental problems than to act collectively and issue Climate Stability Bonds.

We saw in chapter 3 that Climate Stability Bonds combine *efficiency* in achieving their targeted objective, with *transparency* about what this goal is, and the goal's *stability* over time. While each attribute is important separately, their combination could transform policymaking. In today's mixed economies most political debate centres not on what people see as desirable outcomes, but on how to achieve them. There is broad consensus that, for example: unemployment and crime rates must come down; housing, education and health services must all be improved, and the environment should be cleaned up. Yet, despite this broad consensus over *objectives*, progress toward their achievement is haphazard and inefficient, largely because there is so much entrenched disagreement about the ways in which these objectives are best achieved.

The Bond principle would take the politics away from such debate. It would channel market incentives into *efficient* ways of achieving whatever outcome is targeted, provided only that these ways are legal and do not conflict with other societal goals. In the future, after the Bonds have been used and accepted, it may be that people will want to widen their scope so that they target what are today nebulous ideas, such as 'happiness' or 'community'. More likely people will want to restrict the objectives targeted by Bonds, and thus the role of government, to easily quantifiable goals that address either the problems of the disadvantaged or those concerns like the environment that are genuinely in the nature of a public good.

Resources are always going to be limited and the Bond concept will not change that. Priorities and choices will always have to be made: under the Bond principle, governments will still have to decide on which social and environmental problems to solve, and on the sums allocated to their solution. In economic theory, and on all the evidence, markets are the best way of allocating scarce resources to achieve prescribed ends. The Bond principle would allow democratic governments and markets to do what each is best at doing—respectively: prescribing ends, and allocating resources to meet these ends.

In the long run the widespread acceptance of the fact that self-interest can be channelled into solving social and environmental problems could have far-reaching implications. Bonds could be issued targeting global problems in addition to that of climate change. Other environmental problems, such as water pollution or the loss of biodiversity, could be made the targets of future Bond issues. Equally, Bonds could target global social scourges, such as malnutrition and war. Globally backed Bonds aimed at these problems could be bought by corrupt governments. Or these governments could be paid by major bondholders to alter their policies. Either way, they would have incentives to modify their behaviour to help achieve these desired outcomes, whether these include ensuring that food supplies reach their own starving citizens, or doing what they can to achieve trans-boundary objectives such as global environmental goals. The Bond concept is likely to be more effective than current aid programmes, because of the inextricable linkage between the reward and the desired outcome.

The introduction of a Bond regime, whether nationally or globally, may be accompanied by operational challenges and problems, not all of which will be anticipated. But these potential problems should not be overstated. Existing laws, careful choice and specification of targeted objectives and perhaps more assurances as to how government will behave would probably circumvent or remedy most of them.

And the question of how well a Bond regime would achieve society's social and environmental goals needs to be considered alongside existing policy-making methods. Currently policymakers, officials or lobbyists can escape or deflect censure because the adverse results of their policies are difficult to relate to their cause. If a Bond issue were to lead to negative effects, the relationship between these effects and their cause would be easier to identify, and deterring such effects would be simpler, than doing so under current activity- or institution- based funding arrangements.

Bonds targeting poverty, malnutrition or deadly conflicts are a long way into the future. The objective of achieving climate stability appears more immediate, and is more likely, at this stage, to gather the necessary political support. Carefully specified Bonds targeting climate change or its negative impacts could be the perfect way of introducing the Bond concept to the wider public.

Efficiency, of course, is not the only attribute of a successful policy instrument. Climate Stability Bonds make transparent not only the goals of climate policy, but also its returns, measured as the increase in climate stability per taxpayer dollar. The Bonds also clarify the inevitable trade-offs that society will have to make between climate stability and other social and environmental goals. In today's complex economies, many people feel alienated from decision making, which to outsiders can appear remote and specialised. Such alienation can express itself in many forms, including apathy, cynicism and even violence. Too often government intervention in social and environmental matters is seen as bureaucratic, intrusive, and unrelated to relevant outcomes. With their focus on identifiable goals, Climate Stability Bonds or the Bond principle in general, may in a small way help to bridge this gap, and encourage public interest and participation in policy formation. That may prove to be as valuable to society as the Bonds' efficiency gains.

Epilogue

The Bond principle has had an unusual fate for an unusual idea. It has been in circulation now for about 13 years, without being adopted anywhere, to my knowledge. But neither has it been dismissed outright. It tends to provoke initial enthusiasm amongst economists and decision makers, but then to be forgotten, as other more pressing issues arise. Robert Shiller, Professor of Economics at Yale University, wrote to me at the end of 1996, praising the idea, saying that it creates "a large interest group for the solution of important problems", and that "the political and other effects of creating such an interest group could be incalculable."

Others have written to me, suggesting that the private sector could issue Social Policy Certificates ('certificates' because 'bonds' connotes government backing), for such socially worthwhile goals as eliminating malnutrition in developing countries. One correspondent wrote to me after the terrorist attacks on the United States on 11 September 2001, speculating on how to use the Bond principle to isolate terrorists. He sent me an article by Hernando DeSoto, the Peruvian economist, who declared that 'it is time for the West to create new policies that inspire governments to harness the entrepreneurial energy that is already humming among the poor…' I will quote my reply in full, as it discusses the possibility that people other than governments issue Bonds whose effects will be felt beyond national boundaries.

"I question the assertion that 'it is time for the West to create new policies that inspire governments to harness the entrepreneurial energy.' It seems to me that to fulfil the aspirations of people in many countries, their (current) governments need to be either bypassed or distracted, rather than co-opted.

"Fortunately, Social Policy Bonds can do these things, and under a Bond regime a government's support, competence or even good intentions need not be assumed. The key is to specify the objective carefully.

"Let us say that as a long-term programme, the US Government, probably with the help of other governments, NGOs and charities in the US and around the world, decides to raise the literacy levels of boys and girls in Middle East and African countries. (Apart from literacy there are other non-controversial objectives, such as basic health and sanitation, which

would promote development and marginalize the extremists, but education is probably one of the most important. The important thing is that any targeted objective has a numerical indicator that correlates very strongly with what we actually want to achieve.) Here we have an objective that cannot be interpreted as cultural imperialism, and that should marginalize the extremists, and only the extremists. Everybody else will support it, or at least not want to be seen as opposing it. The objective serves US interests indirectly, by driving the wedge that DeSoto refers to between the terrorists and the poor. Most of all, this objective serves the interests of the schoolchildren themselves, as well as those of the relevant countries in general. Under the SPB mechanism, governments, NGOs and the charities involved would collectively decide on the exact specification of the literacy objective, and contribute toward the funds needed to redeem the Bonds. They would probably issue Bonds for several countries but I will use the example of Pakistan; a country that has been a recruiting ground for extremists.

"Say that Bonds are issued that would be redeemed only when the literacy in Urdu and numeracy of Pakistani 7-, 11- and 15- year old boys and girls reached very high levels. Success in achieving this objective could be measured by standardised tests of representative but random samples involving hundreds of Pakistani children. Once issued what would happen? The Bonds could be bought by anyone. Say that the Pakistani Government, as (I am assuming) the largest current supplier of literacy-increasing services, decides to buy the Bonds (or is maybe given them by the issuers). It is now in a position to reap financial rewards by doing what it can to increase the literacy of Pakistani schoolchildren. It will do this, not by fiddling the results of literacy tests (which will have to be carried out, or verified by, the Bond issuers), but by channelling resources into expanded, increased, or improved, literacy classes. It may, for example, change the school curriculum to give literacy in colloquial Urdu a higher priority, or it may enforce laws against truancy. Or it could broadcast literacy programmes on television. It would conduct research into the most efficient ways of increasing literacy in its society.

"If at any time others think they can do a better job than the Government, they will be able to buy the Bonds from the Government. Similarly if the Government just does not want to get involved. People and institutions would buy the Bonds instead and work to modify or supplement the Pakistani school system's literacy teaching. These people, either owners of the Bonds themselves, or contracted by bondholders, could lobby the Pakistani Government to, say, give a higher priority to literacy in schools, but they could also develop literacy teaching projects of their own. They might finance literacy classes on TV (originating within Pakistan, or from outside), or set up village schools, or give prizes to the most literate families in villages etc. It would be up to bondholders to decide on those programmes that will give them the highest increase in literacy per unit outlay. The market prices of the Bonds, and their changes, will supply helpful information as to how fast the objective is being achieved, and whether more funds would be required for this long-term project—if so, more Bonds can be issued: see postscript to this message.

"Note that the Bond mechanism would be helped by the support and participation of the Pakistani Government, but it does not rely on such support. However, it may be that the Pakistani Government would issue Bonds itself, with some intellectual [and financial] backing from outsiders. This would be ideal; it would streamline necessary legislative changes etc; the ultimate effect would be to make achievement of the objective quicker and less costly.

"In my book [ie 'Injecting Incentives into the Solution of Social Problems'—see bibliography] I mention briefly the issuing of globally backed Bonds targeting malnutrition or the number of people killed as a result of armed conflict or natural disasters. If very large sums of money went into such global objectives then the composition and character of governments themselves could be determined by such universally desired objectives. One example of what I mean is that a corrupt and malicious government could be bribed into, say, channelling its tax revenue into agricultural development rather than weapons purchases, if it were in the financial interests of its senior members to do so. Or the Bonds could give a financial incentive to well-intentioned people to depose such corrupt governments, though only as a means toward universally desired aims.

"A couple of figures dramatically illustrate what is feasible: in 1999 (the latest year for which data is available) the developed countries gave as bilateral and multilateral aid to agriculture in the third world a total of US$10.7 billion. But they gave to their own farmers (as market price support, value of trade barriers etc) a total of US$357 billion—more than thirty times as much. The latter figure comes from Organisation for Economic Cooperation and Development (OECD), and is a measure not only of budget transfers, but also of transfers from consumers brought about via by trade barriers (including barriers to exports from poor countries). The point is that sums that are relatively small to the rich countries, such as those given as aid to third world agriculture, can have a large positive effect.

"Apart from the Bonds' efficiency advantages, they also have the very big political advantage that they target outcomes, and that most people are in agreement about broad, basic outcomes like basic education and health objectives. The only people who could not be sold on those things are the very people we want to isolate. To be simplistic for a moment, some of the religious extremists can garner support by opposing mixed schools on moral grounds, but I don't think they could get any support if they oppose the *objective* of literate girls."

It may be that we should look to the private sector to initiate deployment of the Bond principle. Government departments are not renowned for their innovatory flair, and they operate within more severe restraints. The Bond principle would be completely new, and there *are* potential pitfalls with the idea, which I have discussed in chapter 4. There are also more specific questions, again raised by my correspondents, such as how Climate Stability Bonds or Social Policy Bonds would be treated for tax purposes, or whether they would be deemed lottery tickets and so contravene the laws in certain jurisdictions. Work done by my correspondents using the Bond concept to promote software development, and raising other relevant questions are cited on the websites listed below.

These questions are all worth asking and investigating, because the advantages of a carefully crafted Bond issue, in my view, far outweigh any potential disadvantages. Yet despite the enthusiasm of my correspondents and myself, the Bond idea remains an idea. Until recently it has attracted little sustained attention. One reason may be that those who are in a position to take it further are simply overburdened with work. Another may be that the Bond principle cannot easily be pigeonholed. To put it simplistically, it uses market, or "right-wing" methods to achieve societal, or "left-wing" goals, so perhaps making it ideologically unsound to individuals and think tanks of either persuasion. Perhaps more important, the idea can be deployed to achieve quite broad objectives—indeed, that is when it is at its best. In an age of increasing specialisation and fragmentation, when non-governmental organisations, professional bodies and interest groups generally have a narrow focus, this may be another reason why it has not so far been applied. I am, though, pleased to be able to say that in July 2001 I worked as a consultant at the OECD, in Paris, writing a paper on applying the Bond principle to the environment (including climate change) and agriculture. Member countries are scheduled to discuss my paper at a meeting in April 2002.

The Bond principle may be especially suited to the environment, whose complexity means that it is difficult to identify and specify the full range of human processes that generate positive or negative externalities, the degree to which they do so, and the relationships between these processes, natural processes and desired outcomes. There are a staggeringly large number of uncertainties in all this, but there are two overriding certainties about the Kyoto Protocol:

1. There is no question that it will be extremely costly and politically divisive to implement, and
Nobody knows whether Kyoto will help bring about a more stable climate, be almost completely ineffectual, or actually destabilise the climate further.

2. Climate Stability Bonds would be a huge improvement on such an expensive and divisive policy approach. In the long run, the colossal savings that could arise from the injecting of market incentives into the achievement of a stable climate may encourage policymakers to apply the Bond principle even more widely.

Bibliography

Investing for the future (September-October 1991), Ronnie Horesh, UK CEED Bulletin No 35, Centre for Economic and Environmental Development, Cambridge, UK. (Presented as Room Document 3 to the December 1994 meeting of the OECD Joint Working Party of the Committee for Agriculture and the Environment Policy Committee.) This is a short introduction to the application of the Bond concept to the environment.

Social Policy Bonds: Injecting market incentives into the solution of social problems (August 1992), Ronnie Horesh, AEU Occasional Papers, University of Cambridge, Cambridge, UK. A longer introduction to the concept and how it can be applied to social problems.

Injecting incentives into the solution of social problems: Social Policy Bonds (September 2000), Ronnie Horesh, Economic Affairs, **20** (3), Institute of Economic Affairs, London, UK. A short article, outlining the concept and its application to social problems. (An edited version of this paper appears in the Annex to this book.)

Injecting incentives into the solution of social and environmental problems: Social Policy Bonds, January 2001, Ronnie Horesh, iUniversity Press, Lincoln, Nebraska, USA. ISBN: 0-595-15374-7. A book of approximately 35000 words describing in detail the concept and its application to social problems. One chapter looks at actual local and national policy targets (from the UK) and discusses their value to the supposed beneficiaries.

Websites

http://www.geocities.com/socialpbonds. 'Social Policy Bonds'. The first three papers listed under 'Bibliography' can be downloaded free of charge from this site, which also contains links to online retailers of the book cited as the remaining bibliographic reference. This website also links to other sites exploring the Bond concept. There is a guestbook where visitors can read or write comments.

http://willware.net:8080/pspcerts.html This is a discussion page by Will Ware, about privatising of the Bond concept.

http://users.rcn.com/wware1/spb-game.html A multi-player on-line simulation game designed to demonstrate the efficacy of Social Policy Bonds in guiding market forces to implement public policy; also by Will Ware.

http://www.openknowledge.org/writing/open-source/scb/ 'The Wall Street Performer Protocol: Using Software Completion Bonds To Fund Open Source Software Development'.

http://www.ipcc.ch/ The IPCC's own website, from which relevant summary texts can be downloaded.

http://www.junkscience.com A site, of sceptical orientation, with links to current articles and papers and archives on many scientific subjects, including climate change.

http://www.vision.net.au/~daly/shame.htm The 'Greenhouse Hall of Shame', by John Daly: an up to date source of examples of pseudo-scientific, or dishonest reporting, of climate change.

Table 1

Implications of changing the number of Climate Stability Bonds issued

Assume that the market values the net cost of achieving climate stability to be $4.5 billion.

	Number of Bonds issued	Redemption value per Bond $ thousand	Therefore max cost of achievement $ billion	Assumed actual cost $ billion	Therefore redemption minus float price, per Bond $ thousand	float price, per Bond $ thousand
Scenario1	5000	1000	5	4.5	900	100
Scenario2	10000	1000	10	4.5	450	550
Scenario 3	50000	1000	50	4.5	90	910

Appendix

Injecting incentives into the solution of social problems: Social Policy Bonds

This is an edited version of a paper outlining the Social Policy Bond concept, which was published in Economic Affairs, *(volume 20, number 3) September-October 1999. Economic Affairs is the quarterly journal of the London-based think tank, the Institute of Economic Affairs.*

Deregulation of western economies and the freer operation of self-interest in the private sector have made many individuals very wealthy indeed. But the less well off have gained little, and many social objectives remain as remote as ever.

I believe that many of these problems persist because they are tackled in ways that do not stimulate or reward self-interest. This is largely because their solution is in the hands of local or central government bodies, whose programmes suffer, in my view, from a fatal flaw that almost guarantees they will be ineffectual and expensive: they reward people for undertaking activities, rather than for delivering desired outcomes.

Social Policy Bonds

My proposal is that a new financial instrument be created that rewards people only when they actually achieve targeted social goals. Social Policy Bonds would be issued by local or national government and auctioned to the highest bidders. Government would undertake to redeem these Bonds for a fixed sum *only when a specified social objective has been achieved*. The Bonds would be freely tradeable after issue, and their market value would rise and fall. With an uncertain redemption date, and because they would not bear interest, Social Policy Bonds would be quite different from conventional government bonds.

What sort of social problems can Social Policy Bonds solve? In principle, any that can be reliably defined and quantified. Key criteria for policy areas within which the Bonds would show the most marked improvement over current programmes are:

1   existing policies have objectives that are unstated, uncosted, obscure or conflicting; and

2   financial rewards to those involved in achieving objectives are uncorrelated to their effectiveness in doing so.

Unfortunately there are many such policy areas, including:

- Crime prevention
- Employment
- Health
- Education
- Air, water or noise pollution

How would the Bonds work? They would create a group of people, bondholders, who have a strong interest in achieving the targeted social objective efficiently, or in paying others to do so. Consider an example. Assume that an urban authority is prepared to spend a maximum of say £10 million to reduce the crime rate within its borders by 50%. It issues one million Bonds that become worth £10 when the crime rate falls below 50% of current levels for a sustained period—say one year. Because the market is likely to see this objective as unlikely to be achieved in the near future, it may value the Bonds when they are floated at as little as £1.00. (This sum would be used by the authority partially to offset the cost of future redemption of the Bonds.) Now, the purchasers of the Bonds hold an asset that could appreciate in value by 900 per cent if a sustained halving of the crime rate is achieved.

Who would buy the Bonds?

Many people would purchase these Bonds with the idea of holding on to them until they could sell them at a profit. These *passive investors* would have no intention of doing anything to reduce crime. They would want to become 'free-riders', hoping to benefit from any increase in the bond price without actually participating in any crime-reducing projects. But the way markets work would limit the opportunities for these passive investors. The more Bonds they collectively own, the more remote the targeted objective becomes, the lower the market price of their Bonds will fall, and the more they stand to lose as the aggregate value of their bond holdings falls. At some point, then, it would become worthwhile for passive investors either to become, or to sell their Bonds to, *active investors*. These people, or institutions, would use their own capital, or borrow on the strength of the redemption value of their Bonds, to initiate or facilitate crime-reduction programmes. Active bondholders would have an incentive to cooperate with each other to help reduce crime, and to do so as cost-effectively as possible.

Rewarding success

Consider some of the measures that bondholders could put into operation:

- encouraging neighbourhood watch schemes;
- encouraging parents to monitor their children's activity more closely;
- subsidising recruitment of unemployed workers;
- complementing police patrols with private security patrols; or
- subsidising widespread use of window locks or burglar alarms.

Many of these activities are, to some extent, undertaken by local bodies or some arm of government nowadays. The crucial difference is that, under a Social Policy Bond regime, people have incentives to seek out and develop those ways of reducing crime that are most cost effective. A police force, a bureaucracy, or an environmental health department, however well-intentioned, is not rewarded in ways that correlate with its success in achieving its objectives—even if these are explicit. But under a Bond regime, the self-interest of bondholders acts so as to encourage those ways of reducing crime that give rate-payers the best return for their outlay. These ways may have been tried before, or tried in different cities, or they may be new and untried. Bondholders would be motivated to seek out, invent and use the most efficient methods for the city whose crime rate is targeted.

Of course, the bondholders need not participate directly in any crime reduction projects. Their role could be one of financing such projects, on the strength of the redemption value of their Bonds, or on the strength of any increase in the value of their Bonds. Their motivation arises from the anticipated supernormal profit arising from early redemption of the Bonds.

Trading the Bonds

Social Policy Bonds, once issued and sold, must be readily tradeable at any time until redemption. This is critical to the operation of the SPB mechanism. Many bond purchasers will want, or need, to sell their Bonds before redemption—which may be a long time in the future. With a secondary market, these holders will be able to realise any capital appreciation experienced by the Bonds. This would give them a greater incentive to purchase the Bonds in the first place.

But there is another important reason for requiring a healthy secondary market in the Bonds: active investors may be able to speed up only one, or a few, of the processes necessary for the targeted objective to be achieved. Once these investors have contributed what they can, and seen the capital value of their Bonds in line with the increased probability of the Bonds' early redemption, they may have no wish to speculate on the speed at which the remaining processes will be carried out. Other groups of active investors, who will have greater expertise in performing these later processes, must be given an incentive to use their expertise to accelerate attainment of the targeted objective. The possible capital appreciation of Bonds bought from previous owners and sold at a still higher price [or redeemed] provides this incentive. The new owners will, if they are successful in these later stages, realise this capital appreciation.

Cascading incentives

Bonds therefore could therefore flow towards those who are most able to help solve the targeted social problem. In fact, though, it is not necessary for there to be any actual flow of Bonds. Large bondholders might simply decide to subcontract out the required work to many different agents, while they themselves hold the Bonds from issue to redemption. The important point is that the bond mechanism ensures that the people who allocate the finance have an incentive to allocate their finance efficiently and to reward successful outcomes, rather than merely to pay people for undertaking an activity. At the limit we can conceive of just one single buyer of all the Bonds. If this buyer were determined to hold on to the Bonds until redemption, then the Bonds would function as a sort of performance-related contract, with the government paying only when the objective has been achieved. The buyer could contract out most, or all, of the work required to achieve the objective, with the incentives given by the Bonds for speedy accomplishment cascading down from the bondholder to those subcontracted to do the work. Regardless of who actually owns the Bonds, the Social Policy Bond mechanism ensures the people who are charged with solving social problems are rewarded for success.

Too large a number of small bondholders would probably do little to help solve certain targeted social problems by themselves. It is likely then that the value of their Bonds would fall until there were aggregation of holdings by people or institutions large enough to initiate effective problem-solving projects. Even these bodies, might not be big enough, on their own, to achieve much without the cooperation of other bondholders. So there would be a powerful incentive for bondholders to *cooperate with each other* to help solve the targeted problem. Aggregation of holdings, and cooperation of bondholders, would stimulate effective problem-solving initiatives.

Definition and operation

For the Social Policy Bond regime to be effective, the targeted objective must be carefully defined, so that its achievement correlates strongly with what society wants to achieve. For instance, numbers of reported crimes could be targeted, if the objective is to achieve a safer urban environment. But this indicator may be unsatisfactory if, for instance, the crime rate becomes so high that people don't bother to report minor assaults or burglaries to the police. A more appropriate indicator might be derived from responses to victim surveys. Remember also that the objective will be a *sustained* lower level of crime.

Once an objective is close to achievement, the issuing body can float a new set of Bonds aimed at maintaining the achieved outcome, or at further improvements. The benefit per unit outlay of a second bond issue is likely to be higher than that of the first issue because, during the lifetime of the first issue, people would have probably developed more efficient methods and systems for solving the targeted social problem.

Advantages of a Social Policy Bond regime

The main advantage of Social Policy Bonds is that, by injecting self-interest into all stages necessary for solving social problems, they would be *more cost-effective* than current, activity-based programmes. For the same government expenditure, therefore, more can be achieved.

Social Policy Bonds also make policy objectives more *transparent*. By focusing on outcomes, rather than activities, social objectives are explicitly identified, while indirect, as well as direct, means of achieving them are encouraged—but only if bondholders think them more efficient. Focusing on identifiable outcomes would encourage constructive participation in the political process, and mean that measures taken to achieve them would be more likely to attract public support.

The Bonds also guarantee *stability* of policy objectives. They could target goals with a necessarily long lead time and bondholders would not be deterred from taking measures to achieve them by fears of a reversal of government policy—or, indeed, a change of government. Also, for the Bonds to be as successful as possible, governments would have to give assurances as to their future behaviour.

Because Social Policy Bonds focus on outcomes, which can be broad, they have *informational advantages* that make it easier to consider tackling problems that would otherwise be addressed only on an ad hoc basis. Priorities for health service funding, for example, are strongly influenced by groups of medical specialists with little incentive or capacity to see improvements in the *general* health of the nation as an objective. So funding of these specialities depends on the strength of their lobby groups. And what is arguably the most efficient way of spending the taxpayer's health pound – preventive medicine – receives derisory funding because it has no powerful lobbyists.

Targeting broad indicators of well-being – life expectancy, infant mortality, disability – would ensure that scarce resources are allocated in ways that would directly achieve *society's* health objectives. It would be up to bondholders to explore the scientific and financial relationships so as to divert, impartially, their funding into those existing or new areas of the health service that would most efficiently use them to achieve the targeted broad outcomes.

More generally, most social problems will require more than a single project or programme for their solution. Social Policy Bonds will encourage and reward the most efficient range of approaches. This occurs because of the nature of the bond mechanism, and requires no selection or supervision by government of the most efficient programme. Government dictates only the objective, not the way of achieving it.

Government and markets

Government spending in Britain today amounts to about 42 per cent of Gross Domestic Product. Much of the debate about this spending centres on its size, rather than its inefficiency. Yet the two are linked: it is hard to voice the case for reducing the size of government when many social problems persist. And these problems persist, I believe, because the government programmes that aim to tackle them reward doing, rather than achieving. People are paid for their time, rather than their efficiency or success. As a result, government programmes are cumbersome and inefficient. Typically they are unresponsive to events and lack ability to adapt to local circumstances. There is no incentive for the people who run them to do so efficiently. Even worse, some programmes have perverse incentives: if a police force, for example, is too successful at cutting crime one year, it may find its budget cut the following year. Or at least, the possibility that that might happen could have some effect on performance.

Social Policy Bonds, on the other hand, would be explicitly focused on outcomes. As such, they would command wider political support than activity-based programmes. And because they inject incentives into all stages necessary for solving social problems, they will be more efficient than current efforts.

Resources are always going to be limited and Social Policy Bonds will not change that. Priorities and choices will always have to be made: under the Bond principle, governments will still decide on which problems to solve, and on the sums allocated to their solution. But democratic governments are good at representing and articulating their people's wishes. Where they are not so successful is in working out the most efficient ways of achieving these goals. This achievement is really a matter of allocating scarce resources. In economic theory, and on all the evidence, markets are the best way of allocating scarce resources to achieve prescribed ends. Social Policy Bonds would allow both governments and markets to do what each is best at doing—respectively: prescribing ends, and allocating resources to meet these ends.

In the long run the widespread acceptance of the fact that self-interest can be channelled into solving social problems could have more far-reaching implications. International, or even global, social or environmental problems, such as malnutrition or climate change, could be made the targets of future bond issues. However, the acceptance of a Social Policy Bond regime, even with the aim of achieving national goals as uncontroversial as lower unemployment, or better health outcomes, may be politically difficult, and must be a gradual process. But the potential benefits should not be ignored. By harnessing market forces in the service of social goals, Social Policy Bonds could, I believe, deliver better social outcomes with a much smaller public sector.

End Notes

1   *Summary for Policymakers: A report of Working Group I of the Intergovernmental Panel on Climate Change*, IPCC, January 2001. (Downloadable from IPCC website.)

2   *Summary for Policymakers: A report of Working Group I of the Intergovernmental Panel on Climate Change*, IPCC, January 2001. (Downloadable from IPCC website.)

3   *Is Bush right?* Bjorn Lomborg, 'Prospect', July 2001.

4   *The Great Promise of the 'Greenhouse Effect'*, Sylvan H Wittwer, 'Consumers Research', June 1997 (pages 19-22).

5   *Greening Earth Mops Up Carbon Dioxide*, Environmental News Service, Washington DC, 5 September 2001.

6   The Kyoto Protocol:
    http://unfccc.int/resource/process/components/response/respkp.html, as at 21 June 2000.

7   Quoted in *'Reassessing Kyoto Agreement: Scientists See Little Environmental Advantage'*, Joby Warrick, 'the Washington Post', 13 February 1998.

8   *The Truth about the Environment*, Bjorn Lomborg, 'The Economist', 4 August 2001.

9   *Why Kyoto will not stop this*, Bjorn Lomborg, 'The Guardian', 17 August 2001.

10 *The Skeptical Environmentalist)*, Bjorn Lomborg, Cambridge University Press, September 2001. Text in quotes from the IPCC Assessment of Global Warming, 2001 (para 9.3.6.6).

11 *Is Bush right?*, Bjorn Lomborg, 'Prospect', July 2001.

12 *Beach bugs make for a cooler world*, Peter Hadfield, 'New Scientist', 12 July 1997 (page 17).

[13] *There's Gold in That Dirty Mess,* 'Newsweek', 27 August 2001.

[14] *Environmental indicators for Agriculture, Volume 3: Methods and Results,* OECD Paris, France (page 278).

[15] At the International Conference on Population and Development (ICPD) held in Cairo, Egypt, in September 1994, delegations from 179 States agreed to make family planning universally available by 2015, or sooner. From http://www.un.org/ecosocdev/geninfo/populatin/icpd.htm.

[16] *Respiration in the balance,* John Grace and Mark Rayment, 'Nature', vol 404, 20 April 2000 (pages 819-20).

[17] *Is Bush right?* Bjorn Lomborg, 'Prospect', July 2001.

[18] IPCC, Working Group III, Third Assessment Report, Chapter 1: Setting the Stage: Climate Change and Sustainable Development, paras 1.4.3.3, headed 'Appropriate lifestyles', IPCC, 2001.

[19] For example, by the (UK) Cheshire County Council in a small, widely distributed pamphlet: *Target 2000,* Cheshire County Council, 1999.

[20] *Paper versus polystyrene: a complex choice,* Martin B Hocking, 'Science' 251, pp 504-5, 1 February 1991.

[21] *Energy and environmental profile analysis of children's disposable and cloth diapers,* Franklin Associates Ltd, Prairie Village, KS, United States, July 1990.

0-595-21164-X

www.ingramcontent.com/pod-product-compliance
Lightning Source LLC
Chambersburg PA
CBHW030839180526
45163CB00004B/1383